WHY HUMANS COOPERATE

Evolution and Cognition
General Editor, Stephen Stich, Rutgers University

Published in the Series

WHY HUMANS COOPERATE

A Cultural and Evolutionary Explanation

Natalie Henrich
Joseph Henrich

OXFORD
UNIVERSITY PRESS

2007

OXFORD

UNIVERSITY PRESS

Oxford University Press, Inc., publishes works that further
Oxford University's objective of excellence
in research, scholarship, and education.

Oxford   New York
Auckland   Cape Town   Dar es Salaam   Hong Kong   Karachi
Kuala Lumpur   Madrid   Melbourne   Mexico City   Nairobi
New Delhi   Shanghai   Taipei   Toronto

With offices in
Argentina   Austria   Brazil   Chile   Czech Republic   France   Greece
Guatemala   Hungary   Italy   Japan   Poland   Portugal   Singapore
South Korea   Switzerland   Thailand   Turkey   Ukraine   Vietnam

Published by Oxford University Press, Inc.
198 Madison Avenue, New York, New York 10016

www.oup.com

Oxford is a registered trademark of Oxford University Press

Library of Congress Cataloging-in-Publication Data
Henrich, Natalie, 1973–
Why humans cooperate : a cultural and evolutionary explanation / Natalie Henrich and
Joseph Henrich.
    p.   cm.
Includes bibliographical references and index.
Contents: Evolution, culture, cooperation and the Chaldeans—
Dual inheritance theory: the evolution of cultural capacities and cultural evolution—
Evolutionary theory and the social psychology of human cooperation—
The Chaldeans: history and the community today—
Family first: kinship explains most cooperative behavior—
Cooperation through reciprocity and reputation—Social norms and prosociality—
Culturally-evolved social norms lead to context-specific-cooperation—
Ethnicity: in-group preferences and cooperation—Cooperative dilemmas in the world today.
ISBN 978-0-19-530068-0; 978-0-19-531423-6 (pbk.)
1. Interpersonal relations—Case studies.   2. Chaldean Catholics—Michigan—Detroit Region.
I. Henrich, Joseph Patrick.   II. Title.
HM 1106.H44 2007
305.6′815—dc22   2006048326

9 8 7 6 5 4 3 2 1

Printed in the United States of America
on acid-free paper

To our daughter, Zoey Diane. And in memory of Diane Margaret Henrich.

To the Chaldeans, who warmly opened their homes and lives to us.

Acknowledgments

Many people have made this book possible. First and foremost, our thanks go out to the Chaldeans in metro Detroit who welcomed Natalie into their lives with openness and warmth.

This book grew out of Natalie's dissertation, and both the ethnographic and theoretical work benefited from the guidance that Allen Johnson and Robert Boyd provided Natalie while she was conducting her research and writing her dissertation. And as mentors to both of us, their influences can be seen throughout the book.

We also thank Amanda Sheres and Nick Ellwanger for proofreading the manuscript, and Monique Borgerhoff Mulder and anonymous reviewers for their comments. Any errors are ours.

Research with the Chaldeans was funded by grants provided by the National Science Foundation and through the MacArthur Foundation.

Contents

Abbreviations

CC	Cooperation with Chaldeans
CNC	Cooperation with non-Chaldeans
DG	Dictator Game
EI	Ethnic Index
IMO	Income Maximizing Offer
MU	Monetary Unit
MW	Mann-Whitney test
PD	Prisoner's Dilemma
PGG	Public Good's Game
SMUG	Strategy Method Ultimatum Game
TFT	Tit-for-tat
UG	Ultimatum Game

WHY HUMANS COOPERATE

1

Evolution, Culture, Cooperation, and the Chaldeans

A man who was not impelled by any deep, instinctive feeling, to sacrifice his life for the good of others, yet was roused to such actions by a sense of glory, would by his example excite the same wish for glory in other men, and would strengthen by exercise the noble feeling of admiration. He might thus do far more good to his tribe than by begetting offspring with a tendency to inherit his own high character.

—Charles Darwin, "On the Development of the Intellectual and Moral Faculties," chap. 5 of *The Descent of Man*

This book is an inquiry into one of the great puzzles in the human sciences: the evolution of cooperation and altruism in the human species. Unlike many works on this question, we seek to simultaneously draw together formal theoretical work on the evolution of cooperation, rich ethnographic descriptions of human social life, and a wide range of experimental results from both the laboratory and the field. On the theoretical front, we provide an introduction to the puzzle of human cooperation and a unified theoretical framework that integrates culture, psychology, and evolution in a manner that makes these concepts accessible to nonspecialists. From this general framework a set of theoretical foci emerge to provide a backbone for the book and place this work in a context that will evoke a sense of familiarity for those who have studied human societies. These foci are (1) kinship, (2) reciprocity and reputation, (3) social norms, and (4) ethnicity. Though we have attempted to present this theoretical material at an introductory level, we believe that more advanced students and scholars will find our synthesis of culture, coevolution, and cooperation worthwhile and provocative.

Breathing life into the deductive logic and equations that buttress our theoretical presentation, we develop a body of ethnographic material on social life from 18 months of study by Natalie Smith Henrich among the Chaldeans of metro Detroit, as they lived, worked, and socialized at the end of the twentieth century. The community studied in this book is made up of first-, second-, and third-generation Catholic immigrants from Iraq who are predominantly middle- and upper-class inhabitants of Southfield, Michigan. As a focused,

3

problem-driven ethnography, this book uses the lens of evolutionary theory to examine the anatomy of cooperation in Chaldean life. Though we make no attempt to comprehensively explore their lifeways, readers will learn a fair amount about contemporary Chaldean life in metro Detroit in working their way through the book. After laying the theoretical foundations of our inquiry in chapters 2 and 3, we open the ethnographic stream in chapter 4 by providing a brief introduction to Chaldean culture and history, with an eye on issues relevant to our work among the Michigan population.

In general, we feel that the use of both experimental and ethnographic methods is crucial for advancing theory. Experiments allow for precise measures, controlled comparisons, and specific manipulations, which often permit the direct testing of alternative theories. However, it is often unclear whether what is measured and manipulated in experiments actually affects decisions and behavior in real life. In contrast, ethnography, although messier than experiments and only occasionally amenable to controlled tests, allows us to systematically look for a consistency between real life and theory. Experiments on decision-making, for example, allow students using computers in a university laboratory to play the role of hypothetical farmers. Given full information on their screens about prices, yields, and labor costs related to their "crop choices," students make systematic trade-offs among the three variables in deciding "what to plant." Ethnography, however, has shown that many real farmers don't have any idea what the yields and prices are of the crops they don't plant—thus, they cannot be making the trade-offs seen in the laboratory. The experiments are missing the dynamics driving real behavior and decision making (Henrich 2002). To address the inadequacy of relying entirely on either experiments or ethnography, we have used experiments throughout this book to test one theoretical idea or another, and then employed ethnographic data to examine the consistency between laboratory-tested theories and real life.

Specifically, we have integrated experimental data with our ethnographic work in two ways. First, at the beginning of each of the relevant chapters, we summarize the most important experimental findings—predominately from behavioral economics—vis-à-vis the pertinent evolutionary theories. Second, we report results from our own use of experimental tools among the Chaldeans. As we show in chapters 5–9, and have already shown elsewhere (Henrich et al. 2004; Henrich and Smith 2004), the combination of experimental and ethnographic data can be more powerful than either is alone. Thus, in addition to the findings relevant to our focus herein, we hope to persuade experimentalists to read more ethnography and history, and more ethnographers to equip themselves with the rich set of experimental tools available in economics and psychology.

In our emphasis on culture and culture-gene coevolution, this book is unlike most other works in the field of evolution and human behavior. Broadly following the research programs laid out by Charles Darwin (1981) and James Mark Baldwin (1896a; 1896b; 1968), our theoretical approach brings "culture" (as socially learned behaviors, beliefs, values, etc.) into evolutionary theory (Richerson and Boyd 2005). In contrast to noncultural evolutionary approaches, such as those typically found under the rubric of "evolutionary psychology" (Tooby and Cosmides 1992), our framework adds two interrelated elements that we believe are crucial for understanding humans. First, we draw ideas from a large and growing body of evolutionary theory that views our "evolved minds" as a result (at least in part) of the coevolutionary interaction between genes and culture (Boyd and Richerson 1985; Henrich 2004; Henrich and McElreath 2003; Laland 2000; A. Rogers 1989). Given the massive reliance that humans place on social learning (i.e., culture: Tomasello 1999; Tomasello 2000) and the tendency for these socially transmitted behaviors to alter both our local physical and social environments (Durham 1991), it is quite difficult to imagine how cultural transmission could not have affected genetic evolution.[1] Second, the unique nature of human cultural transmission creates chains of learning that operate over successive generations. That is, our cultural capacity gives rise to cultural evolution and suggests that some important aspects of human behavior cannot be understood without considering the cultural evolutionary history of the particular social group in question.

In the coming chapters, we argue that both elements of this dual inheritance approach, culture-gene coevolution and culture history, are of great importance for understanding cooperation in our species and can be effectively incorporated under the umbrella of Darwinian theory (Boyd and Richerson 2002; Henrich et al. 2003; Richerson and Boyd 1998; Richerson, Boyd, and Henrich 2003). Thus, our theoretical perspective allows researchers to consider and develop explanations that involve both genetically evolved psychologies and cultural transmission (and culturally evolved psychologies: Levinson 2003; Nisbett 2003) without manifesting the very popular, false, and intellectually destructive dichotomy between evolutionary and cultural explanations. As we hope to show with regard to the Chaldeans, fully incorporating the Darwin-Baldwin vision of culture-gene coevolution expands our abilities to explain human behavior without losing the critical linkages between humans and the rest of the natural world that evolutionary theory provides.

As an urban ethnography, and especially one that is driven by evolutionary theory, this book is either unique or rare. Evolutionary anthropologists have often been accused of focusing their research on "small-scale," "tribal," or "primitive" human social groups at the exclusion of people in industrialized societies. Moreover, when anthropologists do study populations in an urban context, we tend to either study impoverished groups, marginalized groups, or university students. Here, we offer an urban ethnography of a fairly wealthy ethnic group in a major U.S. city. Thus, we demonstrate that evolutionary theory is as useful in explaining the behavior of middle- and upper-class Americans in an urban environment as it is to understanding people living in small-scale societies.

The Organization of the Book

This book is organized as follows: chapter 2 sketches the broad theoretical background by introducing culture-gene coevolutionary theory, explaining how culture and cultural evolution can be incorporated into evolutionary theory, and detailing a few key aspects of human cultural learning. Chapter 3 narrows our theoretical focus and charts the course for the rest of the book by applying both evolutionary and coevolutionary theory to the puzzle of human cooperation. Chapter 4 sets the scene for our ethnographic discussions by introducing the Chaldeans, their history, some ethnographic background on their life in metro Detroit, and a description of the ethnographic field methods used. Chapters 5 through 9 further develops each of the major classes of theories of cooperation introduced in chapter 2 by supplying further discussion, reviewing relevant experimental work, and examining each vis-à-vis Chaldean social life and cooperation. Specifically, chapter 5 deals with kinship, chapter 6 with reciprocity and reputation, chapters 7 and 8 with social norms and punishment, and chapter 9 with ethnicity. Chapter 10 concludes by considering the cooperative dilemmas associated with some of the environmental, economic, and public health problems that currently confront our society, both locally and globally. The appendices provide more in-depth background material that may interest some readers and are referenced at various points through the book.[2]

2

Dual Inheritance Theory
The Evolution of Cultural Capacities and Cultural Evolution

> The tendency to imitate may come into direct conflict with the prudential teachings of pleasure and pain, and yet may be acted upon. A child may do, and keep on doing, imitations which cause him pain.
>
> —James Mark Baldwin, chap. 10 of *Mental Development in the Child and the Race*

> With the exception of the instinct of self-preservation, the propensity for emulation is probably the strongest and most alert and persistent of the economic motives proper.
>
> —Thorstein Veblen, chap. 5 of *The Theory of the Leisure Class*

Since the rise of human sociobiology in the 1970s, culture and biology or cultural explanations and evolutionary explanations have often been opposed, and the seeming opposition between the categories has led to a great deal of unnecessary dispute and debate. This dichotomy and its associated debates are now outmoded and unproductive. A wide range of human behaviors that most people would think of as purely cultural (dress, greetings, food taboos, etc.) are actually 100 percent cultural and 100 percent genetic. Many behaviors are cultural in that they are socially learned by observation and interaction in a social group—social learning can then be understood as the foundational capacity that underpins what is typically glossed as "culture." All culturally acquired behaviors, beliefs, preferences, strategies, and practices (hereafter, we refer to all these collectively as "cultural traits") are also genetic in the sense that their acquisition requires brain machinery that allows for substantial amounts of complex, high-fidelity social learning. We know that there must be "human genes" that somehow allow for culturally acquired behavior, as chimpanzees reared (enculturated) alongside human children do not acquire anything approaching adult human behavioral patterns or social norms. Interestingly, although human-reared chimpanzees seem to acquire little from their human

7

families via imitation, these families' human children have been observed to readily acquire a number of behaviors from their physically more advanced chimpanzee "siblings," including knuckle walking (even after achieving full bipedality), shoe-chewing, a habit of scraping their teeth against interior walls, excessive biting, and a range of stereotypical chimpanzee food grunts and hoots.[1] More generally, although limited social learning abilities are found elsewhere in nature, social learning in our species is high fidelity, frequent, internally motivated, often unconscious, and broadly applicable, with humans learning everything from motor patterns to goals and affective responses, in domains ranging from toolmaking and food preferences to altruism and spatial cognition. If other animals are "cultural," then we are a hypercultural species (Henrich 2003).

Capacities for social learning—and culture—quite naturally emerge from a Darwinian framework once one sees the dualistic opposition between culture and biology for what it is. In their now classic treatise, *Culture and the Evolutionary Process*, Boyd and Richerson (1985) formally apply the logic of natural selection to the evolution of our capacities for social learning (culture). This work, and many subsequent papers, showed that social learning could be effectively understood as a genetically evolved adaptation for acquiring (learning) adaptive traits in complex, spatially variable, and fluctuating environments. To see this, imagine you were born into a band of hunter-gatherers and needed to figure out what to eat. You could wander around the environment sampling various potential foods until you found some you liked (assume liking is a cue for "good to eat"). This, however, risks eating something poisonous and will certainly waste a lot of time (which could be devoted to other activities), as you are most likely to encounter things that are either nonnutritious or indigestible. A social learner can avoid all of this by simply focusing on the healthiest members of his group and eating whatever they eat. This cultural approach obtains a pretty good diet and avoids the costs of searching, sampling, and possibly eating something poisonous. Similarly, imagine you are again growing up among foragers and need to figure out how to hunt. If you could not learn via imitation, you would have to individually figure out a huge amount of information about animal behavior, tracking, hunting skills, and how to manufacture the necessary technology. Simply making a bow and arrow, which often requires a poison for the arrows to be effective, demands complex skills and knowledge. One would need to figure out, among other things, which trees can be used for bows; how to strengthen and straighten the wood; which insects, larvae, or tree saps can be used for making a poison; and which animal sinews are best for bow strings—all things one is rather unlikely to discover by himself. A good imitator, however, can simply observe and learn from the other members of his group, thereby taking advantage of the accumulated experience and wisdom of previous generations. If people mostly learn from the previous generation but occasionally make additions and improvements through their own experience, experiments, or luck, culture can become an adaptive system of learned traits that accumulate through time. The result of this kind of cultural transmission is a second system of inheritance that, for as long as human ancestors have had sophisticated social learning, has run in parallel—and often intertwined—with our genetic inheritance system.

With the emergence of this cumulative cultural learning it is useful to distinguish three levels of explanation: (1) Ultimate level: natural selection builds the psychological capacities for cultural learning; (2) Intermediate level: culture evolves, accumulates, and adapts nongenetically to produce local skills, preferences, beliefs, values, and cognitive abilities; (3) Proximate level: psychological mechanisms, preferences, values, beliefs, and motivations, which are the joint products of genetic and cultural evolutionary history, propel individual actions, decision-making, and behavior in particular situations.

Emerging from this line of reasoning, Dual Inheritance Theory aims to take account of the evolution of our capacities for culture (sophisticated social learning), cultural evolution itself, and the interaction—or coevolution—of culture and genes over tens of thousands of years. This approach can be summarized with three key ideas:

1. Culture, cultural transmission, and cultural evolution arise from genetically evolved psychological adaptations for acquiring ideas, beliefs, values, practices, mental models, and strategies from other individuals by observation and inference. In the next section, we summarize how evolutionary theory has been used to predict the psychological details of these cultural learning cognitive capacities.

2. These psychological mechanisms for social learning led to behaviors that were, on average, adaptive in the varying ancestral environments of the human lineage. Any particular individual's behavior or group's cultural practices may be adaptively neutral or maladaptive. By specifying some of the psychological details of these cultural learning abilities (see #1 above), cultural evolutionary models enable us to predict the patterns and conditions of maladaptation, and thus provide theories of adaptation, maladaptation, and the dynamic process of cultural change (Boyd and Richerson 1985: chap. 7; Henrich 2004b). This is an advantage over the models traditionally used in sociobiology and human behavioral ecology, in which no proximate mechanisms for achieving adaptation are specified—and thus the dynamics of change cannot be explored, and behaviors are either "adaptive" or inexplicable.

3. The emergence of cultural learning capacities in the human lineage created population processes that changed the selective environments in which genes develop. For example, suppose the practice of cooking meat spread by imitative learning in ancestral human populations. In an environment of 'cooked meat', natural selection may have favored genes that shortened our energetically costly intestines and altered our digestive chemistry. Such a reduction of digestive tissue may have freed energy for more "brain building." In this way, human biology adapted to culturally transmitted behavior. The interaction is called culture-gene coevolution. As discussed below, this interaction may be critical for understanding some aspects of human cooperation, particularly large-scale cooperation

among nonrelatives (Baldwin 1896b; Boyd and Richerson 2002; Boyd and Richerson, 2005; Durham 1991; Henrich 2004a, 2004b; Richerson and Boyd 1998; Richerson and Boyd 2000).

Below, we summarize the theory and especially the evidence for certain key aspects of cultural learning, which will be important to our discussion of the evolution of cooperation. For a more complete understanding of this approach to culture and evolution, readers should begin with Henrich and McElreath (2003) or Richerson and Boyd (2005).

Evolved Psychological Mechanisms for Learning Culture

The approach to understanding culture using evolutionary theory begins by considering what kinds of cognitive learning abilities would have allowed individuals to efficiently and effectively extract adaptive ideas, beliefs, and practices from their social worlds in the changing environments of our hunter-gatherer ancestors (Richerson, Boyd, and Bettinger 2001). This approach diverges from mainstream evolutionary psychology in its emphasis on the *costly information hypothesis* and on the evolution of specialized social learning mechanisms. The costly information hypothesis focuses on the evolutionary trade-offs between acquiring accurate behavioral information at high cost and gleaning less accurate information at low cost. By formally exploring how the costly information hypothesis generates trade-offs in the evolution of our social learning capacities, we can formulate predictive theories about the details of human cultural psychology (Henrich and McElreath 2003). When acquiring information by individual learning is costly, natural selection will favor cultural learning mechanisms that allow individuals to extract adaptive information— strategies, practices, heuristics, and beliefs—from other members of their social group at a lower cost than through alternative individual mechanisms (e.g., trial-and-error learning). Human cognition probably contains numerous heuristics, directed attentional biases, and inferential tendencies that facilitate the acquisition of useful traits from other people.

Such cultural learning mechanisms can be categorized into (1) *content biases* and (2) *context biases*. Content biases, or what Boyd and Richerson (1985) have called *direct biases*, cause us to more readily acquire certain beliefs, ideas, or behaviors because some aspect of their content makes them more appealing (or more likely to be inferred from observation). For example, imagine three practices involving different additives to popcorn: the first involves putting salt on popcorn, the second favors adding sugar, and the third involves sprinkling sawdust on the kernels. Innate content biases that affect cultural transmission will guarantee that sawdust will likely not be a popular popcorn additive in any human society (sawdust is not very tasty). Both salt and sugar have innate content biases for sensible evolutionary reasons: foods with salty or sugary flavors were important sources of scarce nutrients and calories in ancestral human environments. Thus, natural selection favored a bias to acquire a taste for salty and sweet foods so that we would be motivated to obtain and eat them. Of course, if you grew up in a society that only salts its popcorn you may steadfastly adhere to your salting preference even once you find that sugar is the standard popcorn seasoning in other societies. Thus, human food preferences are simultaneously cultural (culturally learned) and innate (influenced by innate content biases).

Content biases may be either reliably developing products of our species-shared genetic heritage (i.e., innate) or culture specific. People may culturally learn beliefs, values, and/or mental models that then act as content biases for other aspects of culture. That is, having acquired a particular idea via cultural transmission, a learner may be more likely to acquire another idea because the two "fit together" in some cognitive or psychological sense.[2] For example, believing that a certain ritual in the spring will increase the crop harvest in the summer may favor the acquisition of a belief that a similar ritual will increase a woman's odds of conception, a healthy pregnancy, and/or successfully delivering a robust infant.

Context biases, on the other hand, exploit cues from the individuals who are being learned from (we will term these individuals "models"), rather than from features of the thing being learned (the cultural trait), to guide social learning. There is a great deal of adaptive information embodied in both who holds ideas and how common the ideas or practices are. For example, because information is costly to acquire, individuals will do better if they preferentially pay attention to, and learn from, people who are highly successful, particularly skilled, and/or well respected. Social learners who selectively learn from those more likely to have adaptive skills (that lead to success) can outcompete those who do not. A large amount of mathematical modeling effort has been expended in exploring the conditions under which different context biases will evolve, how they should be constructed psychologically, and what population patterns will emerge from individuals using such learning mechanisms. Moreover, and perhaps more significant, a vast amount of field and laboratory data confirm that these learning biases are indeed an important part of our cognition (data are sketched below), that they are used by both children and adults, and that they influence economic decisions, opinions, judgments, values (e.g., altruism), eating behavior, rates and methods of suicide, and the diffusion of innovations. Our remaining discussion of psychological mechanisms focuses on the theory and evidence for two categories of context biases in cultural learning: (1) success and prestige biases, and (2) conformity biases.

Success and Prestige Biases

Once an individual is learning from others, she would be wise (in an adaptive evolutionary sense) to be selective about whom she chooses to learn from (Henrich and Gil-White 2001). A learner should use cues, attributes, or characteristics of the individuals in their social world to figure out who is most likely to have useful ideas, beliefs, values, preferences, or strategies that might be gleaned, at least partially, through observation. For example, an aspiring farmer might imitate the strategies and practices of the most skillful, successful, or prestigious farmers who live around him. Simply figuring out who obtains the biggest yields per hectare and copying that person's practices is a lot easier than doing all the trial-and-error learning for the immense variety of decisions a farmer (or anyone else) has to make. On his own, an individual learner would have to experiment with many types of crops, seeds, fertilizers, planting schedules, and various plowing techniques. The variety of combinations creates an explosion of possibilities, making it virtually impossible for an individual to figure out the best farming strategy by relying entirely on experimentation. This is true of many real-world decisions that everyone faces. However, along with figuring out who is the most successful or most skilled, learners should also be concerned about how the things they might learn will fit with their own abilities, the expectations of their role or gender, and their personal context. Learners should assess certain kinds of "similarity" and weigh this alongside their assessments of "skill" and "success." Following this logic, we argue in the next chapter that learners should preferentially learn social norms from individuals who share their ethnic markers (e.g., their dialect, language, or dress; see McElreath, Boyd, and Richerson 2003).

Figuring out who possesses the more adaptive skills, strategies, preferences, and beliefs is no straightforward task. To achieve this, people rely on a range of cues related to skill (or competence), success, and prestige. For rhetorical purposes, this tripartite distinction is helpful because it captures the continuum of cues from direct observation by the learner (of skill or competence) to completely indirect assessments based on prestige. Noting someone's skill or competence, for our purposes, means that one has directly observed their technique or performance. An apprentice might watch two craftsmen working side by side, one hitting all of his marks and gliding right along to a perfect final product (say a handmade chair) as the other struggles, cuts himself twice, curses a bit, and produces something that only the bravest of his friends would sit on.

Cues of success are less direct and take advantage of easily observable correlates of competence (which are hopefully difficult to fake). Depending on the domain and society, such cues might be measured by house size, family size, number of wives and/or children, number of peer-reviewed publications, costliness of car, number of tapirs killed, number of heads taken in raids, size of biggest yam grown, and so on, each of which, in particular social contexts, is related to some domain of skill. Though these cues provide only an indirect measure of competence, they are sometimes more accurate than direct observations of competence. If performances are highly variable, the observations of a small sample of total performance may lead a learner to misperceive competence. However, cues of success often average over many performances, which can help reduce the error in the learner's assessment of whom to learn from.

The evolutionary theory underpinning this form of model-based cultural learning proposes that once the psychological machinery that makes use of competence- and success-based cues for targeted cultural learning has spread through the population, highly skilled and successful individuals will be in high demand, and social learners will need to compete for access to the most skilled models. This created a new selection pressure for learners to pay deference to those they assess as highly skilled (those judged most likely to possess adaptive information) in exchange for preferred access and assistance in learning. Deference benefits may take many forms, including coalitional support, general assistance (helping on laborious projects), public praise, caring for the offspring of the skilled, and gifts (Gurven 2001).

With the spread of deference for highly skilled individuals, natural selection can take advantage of the observable patterns of deference to further save on information-gathering costs. Naive entrants (say immigrants or children), lacking detailed information about the relative skill or success of potential cultural models, may take advantage of the existing pattern of deference by using the amounts and kinds of deference that different models receive as cues of underlying skill. Assessing differences in deference provides a best guess to the skill ranking until more information can be accumulated and integrated. Assessing and learning about whom to learn from using the distribution of deference is merely a way of aggregating the information (opinions) that others have already gleaned about who is a good person to learn from.

In addition to patterns of deference, people unconsciously indicate who they think is a good model through a series of ethological and behavioral phenomena that arise directly from efforts to imitate these individuals. These patterns relate to attention, eye gaze, verbal tones and rhythms, and behavioral postures. As other learners seem keenly attuned to these subtle patterns, it appears that natural selection has favored attention to both patterns of deference and those arising from targeted imitation, as a means of assessing whom to pay attention to for cultural learning. As we will discuss below, a mechanism such as "copy the majority" (conformist transmission) provides an effective way to aggregate the information gathered by observing and listening to others. In this case, conformist transmission can be used to figure out whom to pay attention to for cultural learning.

To understand the difference between cues of prestige, success, and skill, consider the following stylized example of an academic department. A new Ph.D. entering a department and aiming at tenure might assess her senior colleagues in order to figure out whom she should learn from (to achieve her goal of getting tenure). Initially, she can glean a measure of people's prestige by listening to and observing how people act toward one another. If she's really serious, she might pull up everyone's curricula vitae and count their publications (and divide by their "years since Ph.D."). This would give a measure or cue of success. Finally, if our fresh Ph.D. still has not given up all hope of finding a good model, she might read everyone's papers (or at least those who rank high in "success" and "prestige") and watch them teach. This would give our learner a measure of skill. Aggregating all these measures, she'd have a decent estimate of whom to start imitating.

Interestingly, the indirect nature of assessing another person's utility as a cultural model (i.e., a person's possession of adaptive information that is useful to the learner) creates an important phenomenon. In a complex world, such indirect measures do not tell the learner which of the model's behavior, ideas, practices, and strategies causally contribute to his success or competence. For example, are people successful in farming because of what they plant, when they plant, how they plant, or how they make sacrifices to the spirits—or all four? Because of this ambiguity, humans may have evolved the propensity to copy successful individuals across a wide range of cultural traits, only some of which may actually relate to the individuals' success. When information is costly, it turns out that this strategy will be favored by natural selection even though it may allow neutral and even somewhat maladaptive traits to hitchhike along with adaptive cultural traits.

Evidence of Selective Model-Based Cultural Learning

Is there actually any evidence for these learning mechanisms? Yes, huge amounts, and it comes from across the social sciences. The evidence shows that children and adults will preferentially learn all kinds of things from individuals demonstrating particular cues of competence, success, and/or prestige— and there need not be any particular relationship between domains of prestige or competence and the things being learned. Unfortunately, the details don't go much beyond that. For example, we would like to know how different kinds of information are integrated. How important is observed competence compared to prestige? And how important is individually acquired information when it contradicts the behavior of highly successful people? Having looked at a wide range of social learning evidence, it is clear that the tendency to imitate prestigious and successful people is one of the most powerful aspects of cultural learning—a point highlighted by the effect of prestige- and model-based cues (e.g., gender and ethnicity) on the imitation of suicide (which we'll discuss below).

In summarizing some of the evidence for the broad power of success- and prestige-biased cultural learning, we emphasize six main points. First, these imitative patterns spontaneously appear in incentivized circumstances (in which choices influence monetary payoffs or other kinds of returns) and nonincentivized circumstances, and in both nonsocial and social situations, including situations that involve direct competition among the learners.[3] Second, the effects emerge broadly across contexts, including economic decisions, opinions, food preferences, beliefs, styles, dialects, and strategies in situations of conflict. Third, consistent with theory, the amount of cultural learning observed depends critically on the degree of uncertainty found in the environment. As uncertainty increases, so does cultural learning. Fourth, these learning patterns emerge even when the model's domain of competence, success, or prestige is apparently unrelated to the domain in question. Fifth, diverse findings from laboratory experiments in both economics and psychology, using very different experimental paradigms, consistently converge—giving us confidence in their robusticity across contexts. Sixth, the patterns of cultural learning observed in economics and psychology laboratories fit closely with field data—giving us confidence that the effects observed in the artificial context of experiments actually matter in the real world. Below, we first summarize some of the laboratory findings to illustrate points 1 through 5, and then describe a few key field studies that illustrate point 6.

Success and Prestige Biases in Nonsocial Situations   Pingle (1995) confirms that people (well, university students) will imitate the strategies of successful individuals in nonsocial circumstances, especially when payoffs are on the line. Using a series of computerized decision situations, participants had to repeatedly select the amount of three different inputs (e.g., fertilizer, seed, and labor) into a production problem for either 21 or 31 rounds, depending on the treatment. Before each decision—i.e., before setting the final amounts, $(x_1, x_2,$ and $x_3)$ of the three inputs for a given round—subjects could pay to find out what profit they'd get if they used different sets of inputs. In the baseline treatment, subjects could learn only from their own analyses and direct experience (what they earned each round from their chosen inputs). To calculate profit in each round, the subject's inputs were run through a preset production function. This function, which was unknown to players, had only one set of optimal inputs (x_1^*, x_2^*, x_3^*)—these inputs would make the most money for the subjects. In four other treatments, opportunities for imitation were introduced in varying ways and with different costs. Participants in all treatments faced the same environment (the same production function) for rounds 1 to 11 (Block 1). At round 12, the environment shifted and again remained constant through round 21. For treatments 2–4 and the control, there was also a "competitive" environment that commenced in round 22 with an environmental shift and lasted through round 31 (Block 3). During this block, the optimal set of inputs shifted dynamically and depended on what other players had done. This means that participants faced a new environment beginning in rounds 1, 12, and 22.

The different treatments manipulated the information available for imitation: in treatment 1, during each round (starting in round 2), participants could—at a cost—look at the inputs and output of one other subject who had previously played that round. In treatment 2, participants could—at a cost—look at a list of inputs and outputs for that round for all the participants that had gone before them. In treatment 3, before the play for each block commenced, participants were given the best outputs and corresponding inputs of previous players for that block. In treatment 4, each subject watched two other subjects complete all 31 rounds before playing themselves. Each treatment used different participants, and participants were paid according to the profit they earned, which was determined by their choices of inputs.

A comparison of the findings from across the treatments highlights several important points about imitation:

1. Participants use imitation, often to a substantial degree, even when decisions are financially motivated and cost-benefit analysis is possible (but costly). The pattern of results across all four experiments—vis-à-vis the non-imitation control—shows the strength of our propensity for imitation: in round 2 of treatments 1 and 2, which can be compared

directly to round 2 in the no-imitation control, people imitated 87 percent and 57 percent of the time, respectively.

2. Imitation tendencies remain strong even in competitive environments. About 43 percent of subjects imitated in round 22 of treatment 2.

3. People tended to imitate (the inputs of) more successful players (those who got higher outputs). The patterns in the data are explicable only if people are looking at the difference in performance and using that as a cue about when to imitate.

4. Uncertainty causes a substantial increase on the reliance on imitation. In rounds 2, 12, 22, when a new environment is first encountered, rates of imitation are highest.

5. The availability of imitative opportunities, even costly ones, improves the average performance of the group. As a group, subjects in imitation treatments outperformed those of the control.

6. The "imitation environment" affects the average performance of the group. Average performance in treatment 3 and 4 exceeds that of treatment 1 and 2. Treatment 3, in which individuals have the best inputs from previous players, is the only informational environment that avoided a substantial degradation in group performance during the competitive block (Block 3).

Now, let's consider a different incentivized experiment, again from economics. Kroll and Levy (1992) ran two treatments of an investment game. In the control treatment, MBA students had to allocate money among three possible investments, each with different expected returns, return variances, and correlations between returns. With all this known, participants set up their portfolios across the three investments and realized some return each round for 16 rounds. Between rounds, they could adjust their allocations among investments but had no idea what was happening to other players. The "imitation-possible" (IP) treatment was identical except that the allocations and performance of all other subjects were posted (without names or labels) between rounds for all participants to see.[4]

Two findings from this study confirm those discussed above. First, using a round-by-round regression analysis, Kroll and Levy found (to their surprise) that participants strongly mimicked the top performers in the IP treatment. That is, participants used success-biased cultural learning to figure out a "wise" investment strategy. Moreover, a comparison of the control and IP treatments shows that adding the possibility of imitation brought the whole group, even after 16 rounds of direct experience, much closer to the optimal allocations predicted by Portfolio Theory. As in the previous experiment, a competitive environment in which individuals pursue their own interests but are also allowed to freely learn from one another creates a social effect: the group does better.

Interestingly, these imitation mechanisms also can produce maladaptive behavior (i.e., bad decisions). Dual Inheritance Theory shows that cultural learning mechanisms exist because they did better on average than alternative decision-making or learning mechanisms in our ancestral environments (the Paleolithic). Confirming this, Kroll and Levy also note that those who overimitated the top performers in the investment game sometimes ended up losing everything catastrophically because they imitated players who had risked substantial amounts and gotten lucky.

So far, the experiments we've summarized have shown that people will copy other people's economic strategies. But, will they copy their *beliefs* about the state of the world? To test this and to more precisely examine whether people will preferentially copy more successful people under uncertainty, Offerman and Sonnemans (1998) designed the following experiment. To represent an "unknown state of the world," they used an urn with a large number of red and white balls (the variable p equals the probability of picking a red ball from this urn). Players know the urn exists and contains red and white balls, but they don't know how many of each color. They are also told that the number of each type of ball does not change. Each round, four balls are randomly chosen from this urn, which yields five possible conditions (0 red balls, 1 red ball, etc.). Players (students again) earn money in two ways. First, they state their beliefs about the likelihood of each of the different conditions occurring (each must specify five probabilities corresponding to 0 reds, 1 red, . . . 4 reds). Payments are made according to the accuracy of each reported belief compared to the true probability. Second, participants must decide whether to "invest" or "not invest." Not investing leads to receiving a fixed sum for that round. Investing leads to losses if the condition of the round is 0, 1, or 2 reds, but yields substantial monetary gains (about three times the fixed amount for not investing) if three or four reds are picked in that round. This means that, in general, someone who prefers more money ought to generally invest more if they believe p is high.

Three treatments were run for 10 rounds each. In the control treatment, players received no information about what other participants did. They gave one set of beliefs and made an investment decision in each round. In the rounds of treatments 1 (T1) and 2 (T2), players stated their beliefs and investment decisions, then received information about the beliefs, investment decisions, and payoffs of two other players, after which they again stated their beliefs and investment decisions for that round.[5] Treatments 1 and 2 differed in only one respect: in T1, all players faced the same conditions (e.g., 3 red balls and 1 white were drawn this round), so in a sense, everyone received the same information. In T2, each player faced a different independent draw from the same urn each round, so that in some sense, the beliefs and behavior of other players should indirectly contain additional information about the state of the world (p). If people were perfect individual learners, the behavior of others in T1 would not be worth paying attention to, as it does not add any new information. Treatment 2 adds new independent observations except that the learner does not get to observe the condition directly but only via the others' decisions and beliefs.

By comparing players' reported beliefs before and after receiving the information about the beliefs of the others, Offerman and Sonnemans assessed the effects of social information. Interestingly, two different learning biases emerged. First, of the three individuals, those with lower payoffs tended to move toward the beliefs of those with higher payoffs. However, individuals with the highest payoffs (of the three) still moved their beliefs toward the other two. It is as if players adjusted their beliefs by doing a kind of payoff-weighted average of their own beliefs and others'.

Comparing the control with the two treatments reveals a few additional findings that reiterate what we've seen above. First, adding cultural learning opportunities improves the average performance of the group. Players in T1 and T2 made more money than those in the control, and those in T2 did somewhat better than those in T1. Second, the imitation of beliefs occurs in both treatment conditions, although there was more in treatment 2. This is important because a critic might expect individuals to use information as perfect individual learners would (what economists call "rational Bayesians": McKelvey and Page 1990). A perfect individual learner, however, would not imitate in treatment 1 (and it remains a bit murky how exactly a perfect individual learner would extract the necessary information from the beliefs of others in T2). However, the use of other people's beliefs makes good sense from a Dual Inheritance perspective if we assume that human brains are imperfect information processors and that natural selection had to build a brain that could contend with its own imperfection. Culturally learning from others, even if they perceived identical events, can reduce the amount of error that creeps in as information is perceived, processed, organized, and stored.

Social Situations: Success-Biased Imitation in Strategic Conflicts   Maybe people copy other people in situations involving purely individual decisions, but do they really imitate others in situations in which the behavior of others also influences their own payoffs? Yes. Let's begin by reviewing some experiments that have explored imitative learning in competitive market situations called Cournot markets. In these experiments, individuals are put into groups of two to nine individuals (depending on the experiment and treatment). The game is played for some number of rounds (between 20 to 60). Each round, players make one decision regarding the quantity of some product they will produce. As a result, they receive some profit that depends on the quantity they produce and consumer demand, which depends on the aggregate quantity that all players produce (demand goes down as the aggregate production across all the players in a group goes up). All this is calculated using fixed mathematical functions that are unknown to players. At the end, players are paid in real money according to their total profits. Using this basic setup, experimenters have varied the information available and/or the ease of use of that information, or the time available for decision-making. By varying the treatments and using learning models of various kinds (e.g., success-bias imitation, reinforcement, belief-based best-response, etc.) to predict how players change their behavior over the rounds of the game, researchers can study how people adapt to these situations and how different informational conditions draw forth different adaptive learning tools.

To summarize, the results from several studies demonstrate that players adapt by using imitation strategies along with individual learning strategies. When they do imitate, players have a strong preference for other players who have received higher payoffs. In synthesizing the findings of their experimental study, Alpesteguia, Huck, and Oeschssler (2005) write: "If individuals can imitate actions, most of them do so. And some do so almost all the time." These analyses explicitly show that the probability of imitating increases proportionally with the observed difference in payoffs (this finding is remarkably consistent with the evolutionary prediction; see Schlag 1998 and 1999). Interestingly, Offerman, Potters, and Sonnemans (2002), using a similar setup, found that participants used two kinds of cultural learning mechanism: sometimes they imitated the most successful in their trio, and other times they imitated people who chose more group-beneficial options.

The preceding studies are important because decision making is incentivized, and the available information is rigorously controlled. Qualitatively, however, these findings from economics merely confirm older empirical insights from psychology. We will summarize some of the work from psychology to show that these findings are robust across disciplines, experimenters, experimental traditions, and scientific time. Together, results from the two traditions form a potent argument. Not only do these cultural learning mechanisms operate in incentivized decision making, as studied by the economists, but they also appear in nonincentivized situations in which behavior, opinions, and preferences shift both spontaneously and unconsciously.

In testing the early observations of Miller and Dollard (1941) about prestige and social learning, psychologists Rosenbaum and Tucker (1962) used an extremely artificial experimental setup in which pairs of subjects had to pick the winners of horse races. Of course, as with much early work, there were no horses or real money, but only isolation booths, buttons, and colored lights. Although crude, this work did show that "model competence" (or the frequency of correct answers made by a model), strongly affected the subject's propensity to imitate the model's choices (which horse to bet on), even when those answers were unconnected to the imitator's circumstances. Baron (1970), using a similar setup, provided a confirming result. In addition, numerous other studies have also shown how model success biases imitation (Chalmers, Horne, and Rosenbaum 1963; Greenfield and Kuznicki 1975; Kelman 1958; Mausner 1954; Mausner and Bloch 1957). Qualitatively, these studies report basically the same patterns seen above: people copy the choices of successful individuals.

However, people imitate more than choices, decisions, and strategies. Our opinions are deeply influenced by the opinions expressed by prestigious others. About half a century ago, Tannenbaum (1956) investigated the effect of different sources on opinion change and found evidence for prestige learning biases. The sources used were: (1) a prominent individual, (2) a prominent newspaper, and (3) a prominent social group. His 3x3 design tested for how the same source would bias subjects who (1) had prior respect for the source, (2) were previously neutral toward it, or (3) had prior disrespect for it, on an attitude/opinion item they (1) previously agreed with, (2) were previously neutral toward, or (3) previously disagreed with. He tested the effects of the source's positive and negative attitudes and found that subjects' attitudes were pulled closer to the source's even when subjects' prior opinions were contrary.[6] With similar findings, Haiman (1949) also showed that a model's prestige and perceived competence significantly affected the opinions of others, while perceived physical attractiveness, "likeability," and "fair-mindedness" did not.[7]

Experimental data also show that opinions are influenced even when the prestigious individual's domain of prestige is seemingly unrelated to the opinion domain in question. Ryckman et al. (1972) showed that prestigious individuals whose expertise is in the domain in question and those with prestige from other areas of expertise both influence opinion change. In one set of conditions, participants first did an opinion survey on student activism and watched a live speech by either an individual introduced as an expert in student activism or a college sophomore. In each treatment, the same person played the expert and the sophomore and gave the same speech in both roles. After this, participants repeated the opinion survey. In another treatment with the identical setup, participants listened to either an expert on the Ming Dynasty or a college sophomore. In both conditions, people's opinions moved toward those expressed by the prestigious person after hearing the expert but not after hearing the sophomore. In a similar study using the topic of national budget priorities, Ritchie and Phares (1969) obtained similar results with a leading economist versus a college sophomore in the high- versus low-status manipulation.

Same Patterns Are Observed Outside the Laboratory The same patterns observed in contrived laboratory experiments are observed in "real people" (non-student adults) outside the laboratory. One might wonder, then, why laboratory work is necessary at all. The answer is simple: the laboratory allows for tightly controlled methods of isolating cues and of measuring their relative strengths. The real-world data cannot do this as effectively, but it can confirm that what we see in the laboratory does influence real life.

In his massive review of the literature on the diffusion of innovations, Everett Rogers (1995, p. 18) summarizes some of the lessons from 50 years of research as follows:

> Diffusion investigations show that most individuals do not evaluate an innovation on the basis of scientific studies of its consequences, although such objective evaluations are not entirely irrelevant.... Instead, most people depend mainly upon a subjective evaluation of an innovation that is conveyed to them from other individuals like themselves who have previously adopted the innovation. This dependence on the experience of near peers suggests that the heart of the diffusion process consists of the modeling and imitation by potential adopters of their network partners who have adopted previously.

Rogers goes on to devote an entire chapter to explaining how the diffusion of new ideas, technologies, and practices is strongly influenced by "local opinion leaders." Compiling findings from many diffusion studies, Rogers describes these individuals as: (1) locally high in social status (e.g., high status within the village or village cluster), (2) well respected (indicating prestige), (3) widely connected, and (4) effective as social models for others. Rogers's insights are particularly important here because they confirm that success- and prestige-biased cultural learning are important for the spread of novel technologies and practices.

Naturalistic studies using a jaywalking manipulation have consistently found that people preferentially copy the behavior of high-status models. In these naturalistic experiments, high-status models wear business suits, and low-status models appear disheveled and impoverished. In a meta-analysis of seven studies on jaywalking, Mullen et. al. (1990) show that a high-status obedient model increases others' compliance to "no jaywalking" rules, whereas a low-status obedient model has no significant effect.

Studies by sociolinguists show that these processes apply equally well to the transmission of dialect. Labov (1980, 1972) has shown that dialect change is driven by locally prestigious people. In Philadelphia, upper-class working women pioneered novel sound changes, which then spread through the local social strata. In Martha's Vineyard, most folks are not aware of the dialect differences between themselves and mainlanders. Yet they seem to have granted considerable social status to local fishermen—who exemplify the local spirit of resistance and tradition—which has led to inadvertent copying of the salty dialects spoken by these locally prestigious people.

Using the idea of selective model-based cultural learning (context biases), we can even illuminate some of the robust patterns observed in studies of suicide. Data from industrialized societies show that committing suicide and its associated methods are imitated according to prestige and self-similarity (Stack 1987, 1990, 1991, 1996; Wasserman, Stack, and Reeves 1994). For prestige, many studies show that suicide rates spike after celebrity suicides (Kessler, Downey, and Stipp 1988; Stack 1987). This pattern has been observed in the United States (Stack 1990), Germany (Jonas 1992), and Japan (Stack 1996). The individuals who killed themselves after celebrities did tended to match their models on age, sex, and ethnicity, and they even copied the method. In fact, even fictitious suicides are imitated, although real suicides result in effects that are four times greater (Stack 2000). The time trends of these suicides do not show regression to the mean during the subsequent month, so these are not individuals who would have committed suicide in the near future anyway.

Because suicide is strongly influenced by imitation, suicide can spread in epidemic fashion, showing pattern similar to diseases, novel cultural practices, and innovations. In Micronesia (Rubinstein 1983), beginning in 1960 and lasting for at least 25 years, just such an epidemic spread through certain island populations. This case is particularly stark because the suicides were geograph- ically patterned and distinctively stereotyped. The typical victim was a young male between 15 and 24 (modal age of 18) who still lived at home with his parents. After a disagreement with his parents or girlfriend, the victim was visited in a vision by past suicide victims who "called him to them" (we know this from parasuicides). Heeding the call, the victim performed a "lean hanging" from either a standing or sitting position, usually in an abandoned house, until he died of anoxia or was accidentally discovered and saved. In 75 percent of the cases, there was no prior hint of suicide or depression. Moreover, this pattern was restricted to those who could trace their ethnic descent to Truk or the Marshals, and emerged in localized sporadic outbreaks among socially interconnected adolescents (which parallels the U.S. pattern: Bearman 2004), and could sometimes be traced to the precipitating suicide of a prominent son of a wealthy family. This is all the more stark when contrasted with suicide patterns elsewhere in Oceania. For example, the strong gender bias and lean-hanging method contrast with the pattern in Western Samoa, where male-female rates are nearly equal and most suicides (and attempts) involve drinking paraquat, a deadly herbicide (Booth 1999).

Conformity Biases

What do you do when any observable differences in skill, success, and prestige among individuals do not covary with the observable differences in behavior, beliefs, practices, or values? For example, suppose everyone in your village uses blowguns for hunting, except one regular guy who uses a bow and arrow and obtains fairly average hunting returns. Do you adopt the bow or the blowgun? One solution for dealing with such information-poor dilemmas is to copy the behaviors, beliefs, and strategies of the majority (Boyd and Richerson 1985; Henrich and Boyd 1998). Termed conformist transmission, this mechanism allows individuals to aggregate information from the behavior of many individuals. Because these behaviors implicitly contain the effects of each individual's experience and learning efforts, conformist transmission can be the best route to adaptation in information-poor environments. To see this, suppose every individual is given a noisy signal (a piece of information) from the environment about what the best practice is in the current circumstances. This information, for any one individual, might give a 60 percent chance of noticing that blowguns bring back slightly larger returns than bows. Thus, using individual learning alone, individuals will adopt the more efficient hunting practice with probability 0.60. But if an individual samples the behavior of 10 other individuals, and simply adopts the majority behavior, his chances of adopting the superior blowgun technology increase to 75 percent.

The same logic can be applied to the effect of imperfect information about the relative success of others, who may be useful as cultural models. Some individuals may obtain accurate information that allows them to effectively pick out and copy the most successful individuals, whereas others may receive noisy (inaccurate) information about relative success, which prevents them from effectively distinguishing differences. This second group can still take advantage of the more accurate information received by the first group by adopting the traits adopted by the majority. To see this more clearly, imagine a group of 200 individuals, wherein 100 are experienced hunters and 100 are novices who need to figure out which technology to invest in learning. Of the 100 experienced individuals, suppose that 40 use bows and that 60 use blowguns for hunting.

In their current environment (perhaps it recently changed), however, bows obtain more efficient returns, although the difference is small and hunting returns in general are highly variable. Nevertheless, using the returns of the experienced hunters, 40 of the 100 novices selected a bow hunter to learn from, 50 were left confused, and 10 picked a blowgun hunter to learn from (they got bad information because of the noise in returns). Thus 80 hunters now use bows (40+40), and 70 use blowguns. In their confusion, the 50 confused novices decide to use conformist transmission. This will result in more than 53.3 percent of the confused individuals' adopting bows. For example, of the confused 50, 40 might adopt bows, whereas 10 still decide to go with blowguns. After all of the transmission this generation, 120 hunters will use the more adaptive bows, and 80 will use blowguns. If the older (experienced) generation dies, 80 percent of the new generation will use bows.

This kind of verbal reasoning has been rigorously tested in both analytical models (Boyd and Richerson 1985: chap. 7) and extended to more complex environments using evolutionary simulations (Henrich and Boyd 1998; Kameda and Nakanishi 2002).[8] In their computer simulation, Henrich and Boyd investigated the interaction and coevolution of vertical transmission (parent-offspring transmission), individual learning, and conformist transmission in spatially and temporally varying environments. The results confirm that conformist transmission is likely to evolve under a very wide range of conditions. In fact, these results show that the range of conditions that favor conformist transmission are broader than those for vertical transmission alone—suggesting that if true imitation (via vertical transmission) evolves at all, we should also expect to observe a substantial conformist component. This work leads to several specific predictions about human psychology. First, learners will prefer conformist transmission over vertical transmission, assuming it is possible to access a range of cultural models at low cost (which is often but not always the case). Though a direct test of this prediction is lacking, we note that a substantial amount of research in behavioral genetics indicates that parents actually transmit very little culturally to their offspring—once genetic transmission is accounted for, vertical cultural transmission often accounts for less than 5 percent of the variation among individuals (J. Harris 1995, 1998; Plomin, Defries, and McLearn 2001). Second, as the accuracy of information acquired through individual learning decreases, reliance on conformist transmission (over individual learning) will increase. Third, as the proportion of models—in the learner's sample of models—displaying a trait increases, the strength of the conformist effect should increase nonlinearly as well. We address the second and third predictions below.

Consistent with the broad thrust of the above theoretical work, a substantial amount of empirical research shows that people do use conformist transmission in a wide range of circumstances, particularly when problems are complex or difficult to figure out on one's own. This work reveals that humans have two different forms of conformity that operate in different contexts. The first, often called informational conformity, matches the theoretical expectations from models of conformist transmission, and is used to figure out difficult problems.

It results in people actually altering their private opinions and beliefs about something. The second, often called normative conformity, is conformity for the purposes of going along with the group, to avoid appearing deviant. Under this type of conformity, people alter their superficial behavior but often don't change their underlying opinions or beliefs. We argue that the ultimate origins of this second type of conformity can be explained by the evolutionary process that we describe under the rubric of "social norms" in the next chapter and in chapter 7. Below, we sketch some of the evidence for informational conformity (conformist transmission), which is consistent with the predictions derived above, from Dual Inheritance Theory.

Conformist Transmission in Nonsocial Situations

Baron et al. (1996) used a lineup experiment to systematically vary both the difficulty and importance of a task. Working alone, control subjects observed slide projections of a "perpetrator" for some fixed period and then attempted to pick him out of a lineup of four "suspects." Test subjects faced the same task, except each was in a room with two other "subjects" who were actually confederates of the experimenter. To obscure the actual structure of the experiment from the subjects, seven "critical trials" were interspersed with a series of other mundane trials. During these critical trials, both confederates would give the same incorrect answer before the actual subject gave his answer. Conformity was then assessed by comparing the answers of the test subjects on these critical trials to those of the control subjects. Difficulty was varied by changing the amount of time a subject could initially view the perpetrator. In the easy version, subjects viewed the perpetrator for five seconds, and in the more difficult problem they viewed the perpetrator for one-half second. Control subjects selected the correct person from the lineup 97 percent of the time on the easy version and 76 percent of the time on the harder version. Task importance was also varied. In the less important task, subjects performed for the "good of science." In the more important task, subjects performed for the "good of science" and received a lottery ticket in a $20 lottery for each correct answer. Figure 2.1 presents the results.

In Figure 2.1, the y-axis plots the mean number of conforming trials (out of seven possible trials), and the x-axis shows the "less important" versus "more important" (i.e., unpaid versus paid) versions. The dotted line connects the mean number of conforming trials for the treatments when the task was moderately difficult (control group got 76 percent correct), while the solid line shows them when the task was easy (control group got 97 percent correct). The key thing to note is that when importance increased, the social influence decreased as long as the problem remained quite easy. However, when the problem was moderately difficult, social influence increased.

To clarify, in the unpaid situation, subjects were motivated principally by a desire to conform to the two other (confederate) subjects, with only a slight (nonsignificant) effect of using the social information more in the ambiguous circumstance. However, when money was added as a motivation, it overcame

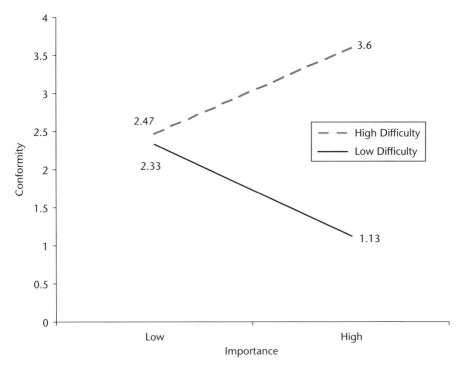

Figure 2.1 Mean number of conforming trials by Condition in the lineup task. Data taken from Baron et al. 1996. ———— Low Difficulty, — — High Difficulty.

much of the desire to conform to the other two (whom the subject did not know) and thereby decreased conformity for the easy problem (where subjects knew the right answer 97 percent of the time based on their own perceptions)—they wanted the money, so they were motivated to give what they were pretty sure was the correct answer. But when subjects were not so sure, but still wanted to get the answer correct, they shifted their reliance toward the choice made by the other two. When an individual's ambiguity about a behavior (or answer) increased and they were motivated to get correct answers, their reliance on social learning increased. This bears out a prediction that comes directly from evolutionary theory.

Using an experimental design that closely parallels the evolutionary simulation constructed by Henrich and Boyd, McElreath, Lubell, et al. (2005) tested the predictions of conformist transmission. Undergraduate subjects repeatedly faced an economic choice between two options, A or B, for 20 rounds. This was posed as a "farming decision" in which A and B were different crops with different yields and variances. Players did not know the mean yields or yield variances for the two crops, but they were told that the local environment might fluctuate such that the mean yields of the crops change. After each round, each player learned the yield realized in that year for his field, and could choose to look at the decisions (crop A or B, but not the yields) of other farmers in the past year. At the end of the 20 rounds, players were paid according to their total yield over the 20 seasons, and made between $4 and $8. Consistent with the theoretical predictions, McElreath, Lubell, et al.'s analysis found that (1) people increase their use of social learning when crop variance is high and decrease it when environmental conditions vary over time; and (2) a simple conformist learning rule (copy the majority) seems to capture an important part of decision making, although there is quite a bit of individual heterogeneity.

A naturalistic experiment using nonincentivized actions further confirms these conformist effects by showing the nonlinear influence of the frequency of a behavior (Coultas 2004). Here, subjects entered a computer lab one by one, not realizing they were in an experiment, and observed a "rare behavior" that involved placing the keyboard cover on top of the monitor. In pretesting, the experimenters confirmed that no one, without modeling, ever put the cover on top of the monitor—so without modeling, the expected frequency of placing the cover on the monitor is zero. The experimenters were able to manipulate the number of individuals placing the cover on their monitor by silently giving explicit instructions to a few through their computers. Others, not receiving these instructions, were observed to see if they placed the cover on top of the monitor. Figure 2.2 summarizes the results by showing how the frequency of models performing the cover placement affected a subject's likelihood of making the same placement. The horizontal axis gives the percentage of individuals already in the room who had their keyboard covers on top of their monitor as the subject entered. The vertical axis gives the probability that the

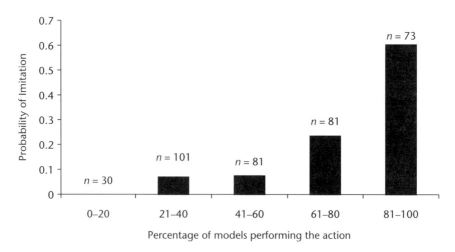

Figure 2.2 The figure shows how the percentage of models performing the "covers on the monitor" behavior influences the likelihood of others performing the same behavior. The n values above each bar gives the number of individuals observed for that bar (e.g., 73 subjects entered a room in which 81 percent to 100 percent of the people in the room had their keyboard covers on their monitors; about 60 percent of these subjects then put their covers on the monitor). The data are taken from Coultas 2004.

subject would then place his keyboard cover onto his monitor. As predicted, the likelihood of performing this behavior, which is not otherwise performed, increases nonlinearly as the percentage of models performing the behavior rises above 50 percent. One problem with this experiment is that it does not carefully distinguish informational from normative conformity.

Conformist Transmission in Social Situations—Particularly in Cooperative Dilemmas

In a study of the effect of social influence on common-pool resource games, Smith and Bell (1994; also see Wit 1999) argue that players sometimes copy other players when they're uncertain about what to do. Two forms of a multi-round common-goods game were used. In both games, one subject and two confederates can withdraw "points" from a common pool that initially contains 15 points. The number of remaining points in the pool doubles every other round, but it cannot exceed 15 points. The game lasts until the common pool goes to zero or until 15 rounds are played. In version 1 of the game, players receive lottery tickets (a chance of winning real money) according to their personal point totals at the end of the game; in version 2, players receive lottery tickets according to their group's point total at the end of the game. In both versions, subjects show a significant reliance on mimicking confederates' behavior. When confederates underutilize the resources, subjects tend to underutilize them as well, and when confederates overutilize, subjects also tend to overutilize resources. Because subjects behave similarly when their self-interest equates with the group's interest and when it's opposed to the group, the authors argue that they mimic as a means of using social information under uncertainty and not as a means of competing.

Broadening this finding, Carpenter also finds evidence of conformist transmission in the behavior of players in a cooperative social dilemma (2004). Similarly, Denant-Boemont, Masclet, and Noussair (2005) observe a conformist effect on the punishment of nonpunishers in cooperative dilemmas where subjects can pay to punish (take money away from) those who don't cooperate, and the subjects even punish those who don't punish noncooperators.

These results, which show that people copy others, especially when the right thing to do is unclear when many others are doing the same thing, are consistent with the predictions of Dual Inheritance Theory and are important because they address economic decisions in social circumstances.

Cultural Learning of Altruism and Punishment

Faced with the preceding evidence for cultural learning, a skeptic of the importance of culture for understanding social behavior might argue that although the predicted effects are observed in the social learning strategies found in laboratory experiments and in the costless expression of opinions or dialects, they probably do not influence the acquisition of durable values or preferences that favor taking individually costly actions, such as those related to altruism or cooperation. We take this challenge seriously in this section by offering converging lines of evidence that demonstrate that people do use cultural learning to acquire intrinsic motivations (preferences or values) associated with at least one kind of altruism/cooperation.

Psychologists, mostly during the 1960s and 1970s, have intensively studied the imitation of a particular type of altruism (i.e., helping strangers in a nonrepeated context) in children. In the paradigmatic experimental setup, from which endless variations emerged, a child is brought alone to the experimental area (often a trailer on school grounds) to get acquainted with the experimenter. Then, the child is introduced to a bowling game and shown a range of attractive prizes that he or she can obtain with the tokens (or pennies or gift certificates) won during the bowling game. Subjects are also shown a charity jar for poor children where they can put some of their winnings if they want. This jar is next to the bowling game and often has a "March of Dimes" poster over it, or some facsimile. A model, who could be a young adult or another peer, demonstrates the game by playing 10 or 20 rounds. On winning rounds, which are preset, the model donates (or not, depending on the treatment) to the charity jar. After finishing the demonstration, the model departs (or not, depending on the treatment), and the child is left to play the bowling game alone, often monitored through a one-way mirror. Putting money in the jar results in the child getting fewer toys or other desirable items, so this is clearly costly altruism (and postgame interviews with subjects confirm that they understood the costs).

The results from numerous researchers involving hundreds of children (ages 6 to 11) and a wide range of setup variations demonstrate three robust findings for our purposes here. First, children spontaneously imitate either the generosity or selfishness of the model. That is, if the children see a model donate to the charity jar, they will donate more, whereas if they see a model fail to donate, they will donate less (than they would without any model). The more the model donates, the more the children donate, and the more opportunities a child has to observe the model, the greater the degree of imitation (Bryan 1971; Bryan and Walbek 1970b; Grusec 1971; Presbie and Coiteux 1971). Similar patterns occur whether models are peers or adults (Hartup and Coates 1967). Second, beyond merely imitating the practice and the amount, children also will readily imitate other aspects of the model's behavior. For example, if multiple charity jars are used (with different recipients), children are influenced by the model's distributional preferences (Harris 1970, 1971). Similarly, Bryan (1971) showed that children not only imitate the amount of generosity, but also the order of putting money in the charity jar versus their personal jar.[9] Third, the effects of exposure to a model endure over months in retests (without a model present) and extend to somewhat similar contexts (Elliot and Vasta 1970; Midlarsky and Bryan 1972; Rice and Grusec 1975; Rushton 1975).

Showing the power of cultural learning, the modeling of altruism in experimental situations such as this has consistently, and often to a substantial degree, emerged as the most effective means to instill altruistic giving in children. Several studies compare the effect and interaction of models that, on the one hand, practice generosity or selfishness and, on the other, preach ("exhort") either generosity or selfishness (e.g., "one ought to donate...."). Preaching alone has been shown to have little effect (or sometimes a negative effect; see Grusec, Saas-Kortsaak, and Simutis 1978). However, even when it has been observed to increase donating, it is always overshadowed by modeling effects. That is, children tend to ignore exhortation toward costly actions when those exhortations are inconsistent with the model's deeds (Bryan, Redfield, and Mader 1971; Bryan and Walbek 1970a, 1970b; Rice and Grusec 1975; Rushton 1975).[10] In postexperimental recall tests of what models said (preached) and did (donated), subjects had the most difficulty in accurately remembering the words and deeds of "inconsistent" models (Bryan and Walbek 1970a).

Figure 2.3 summarizes some experimental results that usefully illustrate several of the above points. Here, children were tested in the basic bowling experiment using six different treatments and were retested six months later. Along with a no-model control condition, children observed a model who (1) preached generosity, (2) preached selfishness, or (3) merely showed positive affect ("This is fun"). This preaching condition was paired with donating either (1) generously or (2) very little. The vertical axis gives the fraction of tokens the subjects donated to charity (to "poor children"), with the horizontal axis separating the treatment groups. This figure illustrates that (1) modeling

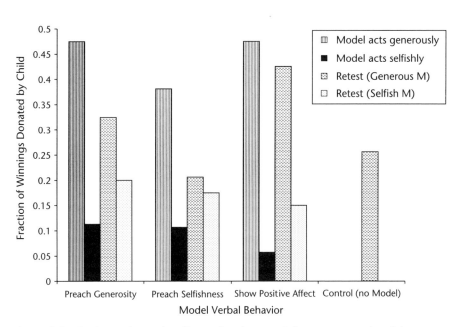

Figure 2.3 The figure shows the effects of various modeling treatments involving combinations of demonstrated behavior and verbal preaching, measured both immediately after modeling and on a retest two months later. The vertical axis shows the fraction of the subject's winnings donated to help "poor children." Data gleaned from Rushton 1975.

donations has the biggest effect on children (note the vertically striped bars across preaching conditions); (2) modeling is augmented by showing either positive affect or preaching generosity (compare the vertically striped bars for "positive affect" and "preaching generosity" with "preaching selfishness"); (3) this effect endures on retests two month later; and (4) showing positive affect increases the effect of imitation more than either type of preaching (compare the difference between the vertically striped and black bars across conditions, and in the retests).

Finally, using the same type of bowling game setup, psychologists have shown that children will imitate standards for self-reward or self-punishment and readily impose those standards on others. In this work, children observe a model rewarding himself by taking tokens when he gets above a certain score, or punishing himself by returning tokens to a bin when he gets below a certain score (Bandura and Kuper 1964).[11] Having learned via imitation or instruction, these children are then allowed to teach another novice, on whom they spontaneously impose the same standards that they recently acquired. These effects endure in retests four weeks later (Mischel and Liebert 1966).

Of course, none of this is limited to children. Both experimental and naturalistic studies show that adults are influenced by imitation. We saw above that undergraduates with real money on the line readily imitate others in deciding how much to contribute in cooperative dilemmas, both when self-interest conflicts with the group's interest and when self-interested choices correspond to group-interested choices. Additionally, in quite natural settings, providing models for imitation has been shown to increase (1) volunteering to help in experiments (Rosenbaum and Blake 1955; Schachter and Hall 1952), (2) helping a stranded motorist, (3) donating to a Salvation Army kettle (Bryan and Test 1967), and (4) giving blood (Rushton and Campbell 1977). Modeling treatments often increase helping rates by 100 percent or more vis-à-vis control treatments.

These empirical findings on imitation and altruism set the stage for the acquisition of social norms via cultural learning, which we'll discuss in the next chapter. Summarizing the above, cultural learning appears to be the most powerful means through which children readily acquire certain kinds of altruistic behaviors (which can be generalized to similar situations) and standards of conduct or performance (including self-rewards or punishments), which they spontaneously apply to peers.

Culture-Gene Coevolution

As just highlighted, the power, fidelity, and importance of cultural learning in our species leads to a second system of inheritance that, in turn, can influence genetic evolution. In the next chapter, one of the theories we will discuss adds clarity to how cultural evolution, driven by competition among cultural groups, may have altered the selective environment faced by genes. We argue that this coevolutionary interaction, perhaps operating over tens of thousands or even hundreds of thousands of years, has deeply affected human sociality, altruism, and cooperation across the entire species. In chapters 6–9, we provide support for this coevolutionary approach by comparing its empirical entailments with those of alternative theories using a combination of ethnographic and laboratory findings. However, because some readers may not be familiar with the degree to which culture can influence genetic evolution in general, we will first highlight research that shows how cultural evolution has influenced recent genetic evolution (occurring in the last 10,000 years) and patterns of genetic diversity in our species. It is important to realize that in contrast to our later substantive arguments that focus on universal aspects of our coevolved social psychology, the findings summarized below are interesting because they sharply reveal "culture-gene coevolution in action" by taking advantage of the synchronic correlations between certain genes and certain cultural traits.

The classic example of culture-gene coevolution emerges from the historically strong correspondence between the prevalence of "lactose absorption" by adults and certain culturally transmitted practices. As background, we offer a few facts: most mammals and all primates have young that produce an enzyme called lactase, which allows them to process the lactose sugars found in milk (for example, from humans, cows, camels, and goats), and thereby access the available nutrition. The production of the lactase enzyme appears to be a developmental phase necessary for using mother's milk during infancy, and typically disappears well before adulthood. This dominant mammalian and primate pattern allows us to associate the ancestral human state with the inability to absorb milk sugar in adults. Interestingly, in some populations of our species— principally those with descent from northern Europe and nomadic pastoralist groups in Africa—most people possess the ability to process milk sugars through adulthood. The coevolutionary hypothesis that best fits the data goes like this: As populations began adopting the culturally transmitted practice of domesticating large mammals (cows) for meat and perhaps for milk to supplement children's diets, a novel selection pressure was created favoring genes that extend the lactase production into adulthood, thereby allowing adults to reap the benefits of the nutrition available in milk. In some places (northern Europe and pastoral Africa), natural selection eventually responded with a "genetic solution" (Bersaglieri et al. 2004; Mulcare et al. 2004). However, in places such as the Middle East and China, the cultural transmission of the practice of keeping large domesticated animals was followed relatively rapidly by the evolution of the technological know-how for (and practice of) turning milk into cheese and yogurt. In these forms, lactase is not required, and anyone can obtain the nutritional benefits of milk. In these populations, adaptive cultural evolution beat natural selection acting on genes to the punch (and thereby sapped the strength of selection on genes) by taking advantage of the opportunity created by the presence of domesticated mammals. As expected, we find the relevant alleles, lactase enzymes, and milk-drinking practices among populations who, historically, had domesticated animals but had not developed the cheese- and yogurt-making know-how or practices[12] (gleaned from: Bayless and Rosenwei 1966; Beja-Pereira et al. 2003; Bersaglieri, et al.

2004; Cavalli-Sforza 1973; Durham 1991; Holden and Mace 1997; J. Johnson et al. 1977; Simoons 1970).

Our second example comes from the Sino-Tibetan–speaking hill tribes of Thailand. Oota et al. (2001) explored the influence of two different postmarriage residential cultural practices on the genetic diversity of the Y-chromosome (which is transmitted from father to son) and on mitochondrial DNA (mtDNA is transmitted from mothers to their offspring of either sex). The research took DNA from people in three villages with matrilocal residence practices, where married couples take up residence in the wives' village, and compared this to three economically and ecologically similar villages with patrilocal residence practices, where couples live in the husband's village. The researchers predicted *apriori* that if culture influences genetic patterns, then the genetic diversity of mtDNA would be relatively greater in the patrilocal villages while the diversity on the Y-chromosome DNA would be greater in the matrilocal villages. Their findings strongly confirm the coevolutionary hypothesis.

These examples illustrate that, even on shorter evolutionary time scales (within the last 6,000 years for lactase), cultural evolution can powerfully influence genetic evolution. Unfortunately, evolutionary approaches to humans have been slow to recognize this and incorporate it into theory.

Summary and Onward to the Puzzle of Cooperation

Our goals in this chapter were threefold. First, we wanted to show that culture and cultural explanations have been incorporated into a fully evolutionary approach, thereby extinguishing the misleading opposition between "biological" and "cultural" explanations that has plagued much thought on human behavior. Second, we sought to demonstrate that evolutionary theory has been fruitfully applied to understanding the psychological foundations of cultural learning, showing that key aspects of cultural learning (e.g., attention and memory) can be understood as a set of specialized psychological adaptations (e.g., prestige and conformist transmission) designed to extract a vast array of useful information from other individuals. Here, we spent substantial space highlighting some important lines of evidence from across the social sciences, both from the field and laboratory, to show that these cultural learning mechanisms are empirically well grounded and influence the real world in significant ways. So, even if one is dubious of our evolutionary explanations for these cultural learning mechanisms, there remain the empirical facts of the operation of cultural learning in the world. Third, we sought to illustrate how our capacities for cultural learning, which are genetically evolved, give rise to a second system of cultural inheritance that has also shaped our genome.

In the next chapter, with each of the above points as background, we will lay out five theories for the evolution of cooperation in humans related to kinship, reciprocity, reputation, social norms, and ethnicity. We argue that both purely genetic and culture-gene interactions have shaped human social psychology such that people cooperate, help, trust, and punish in highly patterned and often contextually specific ways. Chapters 5 through 9 accomplish two things: first, we bring laboratory evidence from behavioral economics and other sources to test and refine these theories; second, we look for the expected social and psychological patterns in Chaldean social life using the theoretical lenses constructed in the next chapter and grounded in the laboratory.

3

Evolutionary Theory and the Social Psychology of Human Cooperation

How selfish so ever man may be supposed, there are evidently some principles in his nature, which interest him in the fortunes of others, and render their happiness necessary to him, though he derives nothing from it, except the pleasure of seeing it.

—Adam Smith, *Theory of Moral Sentiments*, 2000 [1759]

Although a high standard of morality gives but a slight or no advantage to each individual man and his children over the other men of the same tribe,... an advancement in the standard of morality will certainly give an immense advantage to one tribe over another. A tribe including many members who, from possessing in a high degree the spirit of patriotism, fidelity, obedience, courage, and sympathy, were always ready to aid one another, and to sacrifice themselves for the common good, would be victorious over most other tribes; and this would be natural selection. At all times throughout the world tribes have supplanted other tribes; and as morality is one important element in their success, the standard of morality and the number of well-endowed men will thus everywhere tend to rise and increase.

—Charles Darwin "On the Development of the Intellectual and Moral Faculties," chap. 5 of *The Descent of Man*, 1871

The *Why* Puzzle

All around us we see people contributing to the welfare of others, even when it is not convenient and may be costly in terms of time or money, or may affect their personal or professional relationships. In fact, we see so much of this sort of altruism in daily life that we usually don't notice it or stop to question why people are bothering to help others, or how such seemingly ubiquitous generosity might be explained. When people are asked why they help others, a common reply is that doing so is "the right thing to do" and that people "should" help one another. Some scholars have merely accepted such acts as part of being human (Durkheim 1933), without endeavoring to question why we help sometimes, but not at other times, or why different societies seem to provide help to differing degrees and in different domains. In fact, not only are there times that we don't help when we know we could have, there are many times when we don't even perceive an opportunity for helping when, in fact, one exists. Taking this commonsense observation as a point of departure, we will address the question of when and why people incur personal costs in order to help another person or group of people.

Even casual observation suggests some robust patterns in this kind of social behavior. You've probably noticed that people can be quite particular about whom they will help, when, and how much. First, people act frequently, and sometimes at great cost to themselves, to help their families, and especially their children. Why is that? People also help friends, and sometimes acquaintances, but there is something different about "the rules" for helping these people versus helping close family. Friends who break the helping rules often drop from "friend" to "acquaintance." Children, on the other hand (even as adults), are not only given more latitude, but are evaluated using quite different rules. Why are friends different from family members, and where do these "rules" come from? Moreover, what about helping strangers? For instance, have you ever considered why you would willingly stop to give directions to a lost visitor on the street rather than continuing onward? Such an action wastes your time, has some risks (the person may be a thief, con artist, or murderer), and you'll likely never see that person again. If you don't stop to help this stranger, why would you feel bad about it? Are you more likely to help some strangers than others? If so, whom are you more likely to help, and why?

Addressing such why-questions requires distinguishing at least two levels of explanation. The first, the proximate psychological level, focuses on understanding the psychological processes and preferences that propel certain decisions and behaviors. For example, how does the psychology associated with loving your kin work? Who qualifies as kin? Are there different kinds of kin who get different amounts of love or help? How do emotions, affect, and cognition jointly produce behavior and explain variation? The second level, the ultimate evolutionary level, explores the evolutionary processes that produced the proximate psychologies that, in turn, produce decisions and behaviors. However, because we are applying Dual Inheritance Theory, we need to partition the ultimate level into subprocesses to highlight how cultural and genetic evolution jointly influence human behavioral patterns and psychology. Sometimes cultural evolutionary processes will produce psychological and behavioral change and adaptation by altering skills, know-how, preferences, affective responses, and cognitive abilities. All the while, genetic evolution stands dimly in the background, having generated the cultural learning machinery (as explained in the last chapter) and the basic building blocks available to culture (hands, eyes, motor coordination, imitative abilities, anger, etc.) but not contributing more directly to the ultimate explanation. Other times, especially when the relevant psychological adaptations began evolving prior to the emergence of cultural capacities in the human lineage, genetic evolution will have played a direct role in building the relevant psychological and behavioral patterns. At this point in human evolutionary history most aspects of the proximate psychological mechanism that drive human behavior are, in some fashion, a joint product of our intertwined inheritance systems.

This chapter lays out a set of evolutionary theories, derived from a single framework, aimed at explaining the ultimate origins of different aspects of cooperative or altruistic behavior, of which the instances of giving help mentioned above are but examples. Applying these ultimate-level theories, we are able to derive predictions and set up research questions about the details of the proximate psychological mechanisms for social behavior and cooperation. Addressing the ultimate why-questions allows us to better address the proximate psychological (and thus behavioral) questions, and confronting the proximate psychological issues, in parallel, yields the only way to fully test our ideas about ultimate causes. To this end, subsequent chapters bring experimental and ethnographic results to bear on these theories, and these theories in turn illuminate the social life of the Chaldeans in metro Detroit.

What We Mean by "Cooperation" and "Prosociality"

Cooperation occurs when an individual incurs a cost to provide a benefit for another person or people. Costs include such things as money, time, labor, and food. Throughout our discussions, we often refer to cooperative acts as "giving help"—but cooperative acts are not limited to giving help. Cooperative interactions take place within pairs, small groups, or large groups, and can occur among friends, relatives, or strangers. In pairs, cooperation might involve babysitting, giving rides, loaning sugar to a neighbor, or making dinner for a sick person. Among large groups, examples of cooperation include voting, participating in Neighborhood Watch, recycling, contributing to public radio or tsunami victims, sharing food, and paying taxes.[1] In these cases, a large group of people benefit from the costly actions of individuals. To fully understand why these behaviors qualify as cooperation, let's look more closely at three of them: voting, food sharing, and recycling.

When a person goes to the voting booth on Election Day, she incurs a cost. There is the time it takes to drive or walk to the voting place, the time it takes to vote, and the time it takes to return to work. (In the United States, unlike many countries, voting occurs on a workday instead of a holiday.) There may be a financial cost if the voter has to pay for parking and/or gas, and there is an opportunity cost insofar as the person could have been doing something else (such as finishing a report or spending time with one's children). Group benefits also accrue from voting, namely the support of the democratic system and the electoral process—if no one voted, the system would collapse, and the government could be deemed illegitimate. Thus, all members of the voting community share this benefit. Contrary to what many voters believe, however, one of the benefits of voting is not that their candidate of choice is more likely to be elected because of their ballot. Even in the historically close 2000 election of George W. Bush, the decisive voting in Florida separated Bush from Gore by more than

537 votes—536 people could have stayed home with no change in the outcome.[2] Although it is true that the democratic system would collapse if no one voted, it is also true that any single vote has a negligible effect. Therefore, when a person votes she performs a costly act that helps the group (i.e., preserves democracy and legitimacy) but does not help herself (since her one vote is negligible). Consequently, voting qualifies as a large-group cooperative act.

Food sharing, also a cooperative act, can be found throughout the world. Within our own society, food sharing includes offering some of your meal to others when eating at a restaurant, providing food and drink to guests in your home, and donating nonperishables to food banks. In hunting-and-gathering societies, food sharing is widespread and important. Among groups such as the Aché of Paraguay (Hill 2002) and the Hadza of Tanzania (Hawkes, O'Connell, and Jones 2001), meat sharing is the norm, with hunters routinely sharing their kill with the group. In sharing game, the hunter incurs a personal cost (his time, effort, and the loss of some of the meat that he and his family could have consumed) while benefiting the rest of the group by providing them with meat, an important source of protein and nutrition. Though the successful hunters pay a cost in sharing, families who consistently contribute little meat still receive an equitable share from successful hunters. Like voting, the distribution of costs and benefits qualifies food sharing as a form of cooperative behavior, which likely has deep evolutionary importance.

Recycling is an excellent example of large-scale cooperation. In the 1990s, vigorous campaigns were launched to promote the recycling of paper, glass, metal, and plastic products. In many North American cities, each household receives boxes for each type of recyclable material, and the cities provide regular pickup of these items. When we recycle, we incur several costs, including washing out cans and bottles rather than tossing the dirty containers directly into the garbage can, sorting our garbage by product type, and taking a multitude of containers to the curb on collection day (assuming we can even remember which day each type of recyclable is collected!). Although this process is not a huge burden, it certainly takes more time and effort than simply throwing all our refuse into the trash, and often requires special containers, more attention and memory, and more household space. Now consider the benefit from these individual-level costs: the planet and all of its inhabitants get to live in a cleaner, healthier world. What happens if you (one person) decide not to recycle? Nothing. The contribution of any one person is insignificant in terms of the planet's health, and you and your kids still get to live in a cleaner world, as long as many other people pay the costs of recycling.[3] Thus, the benefits that you personally create by recycling do not outweigh your costs, and consequently any self-interested, rational individual who weighed the costs and benefits of his actions should refrain from recycling. Yet if everyone did this, our species would suffer in the long run.

Recycling, like voting, is an example of a public goods problem. In a public goods situation, individuals have an incentive to refrain from cooperating even though, in the long run, everyone will suffer from the loss of the joint benefit. In public goods situations, individuals can "free-ride" by not cooperating while still reaping the benefits created by the contributions of others. If you have ever been in a group project in school, you may have experienced free-riding in the form of a classmate who avoided working but still reaped the benefit (the grade) earned by your own and others' work. In the recycling example, a free-rider would be someone who chooses not to recycle, but still enjoys the cleaner environment that results from those who do. Throughout the book we refer to noncooperation as "free-riding," "defecting," or "cheating."

Besides cooperative behaviors in which an individual provides a direct benefit to others at a cost to herself, there is a larger class of behaviors that we will call *prosocial*. The best example of a prosocial behavior is what economists call *altruistic punishment* (Fehr and Gächter 2002). Here, an individual pays a cost to inflict a cost on another individual in order to maintain a behavior, or norm, in a group. For example, in the gasoline crisis of the 1970s, there were long lines at the pumps. Occasionally, individuals would attempt to free-ride by entering the line near the front. Inevitably, this free-riding would infuriate at least one person in the line, who would often threaten or even physically assault the interloper. The presence of the punishers no doubt dissuaded some other individuals from entering the line near the front. This altruistic punishment was costly for the punisher, as he or she (it was usually a he) risked getting beat up, but it benefited the group by maintaining orderly lines.[4] Nonpunishers in the line reap the benefits created by punishers without having to pay the cost of wrestling with frustrated and enraged line-jumpers. As we will discuss in more detail below, theoretical models have repeatedly shown that if individuals are willing to punish others at a cost to themselves, some otherwise puzzling forms of cooperation can be explained. If there are altruistic punishers out there, anyone who can learn will fall into line, even if he is completely selfish. Furthermore, experimental evidence provides quantifiable evidence from multiple cultures of people's willingness to punish anonymous strangers at a cost to themselves (Henrich, et al. 2004). Thus, our term *prosocial* encompasses both cooperation as we described above ("helping") and altruistic punishment. We have avoided extending "cooperation" to cover altruistic punishment because of the lack of fit with common intuitions about the term "cooperation."

The Mystery of Cooperation and Prosociality

As we saw above, cooperation and altruistic punishment always involve a cost to the cooperator or punisher. This led many evolutionary and rational choice theorists to ask the natural question: If cooperation is costly to the individual, why does anyone do it? Evolutionary biologist Richard Dawkins (1976) and others have emphasized the logic through which natural selection produces "selfish genes." The logic suggests that genes which, on average, cause their bearers (the individuals) to pay fewer costs and reap more benefits relative to others are the ones more likely to be transmitted into the next generation.[5] If a person has a gene that leads her to incur costs to help other individuals, then this individual will likely produce fewer surviving offspring than those who do not possess this gene. Taken at face value, this verbal reasoning indicates that cooperation will generally be filtered out over time by natural selection and that cooperation ought to be rare, both in humans and throughout the rest of nature.[6]

Let's illustrate this with a more concrete and ethnographically relevant example. Consider a person with a gene that leads her to engage in sharing food. By sharing food with the group she is increasing the fitness of everyone else in the group by providing them with extra calories and nutrients. However, at the same time, she is lowering her fitness and the fitness of her offspring by taking away food from them and giving it to others in the group. All else being equal, her generosity will result in her rearing fewer healthy offspring to adulthood than a person who avoids sharing. The generous food sharer will likely have some daughters who also share food with the group, and these daughters, like their mother, will have fewer offspring than the nonsharers in their group. In each generation, the frequency of the "sharing genes" will decrease because of the behavior they promote. Even if "food sharing genes" were initially very common, they would gradually disappear from the population over many generations. Nevertheless, we see lots of food sharing in the world.

Though the broad thrust of theoretical evolutionary biology suggests that cooperation and prosociality should be generally rare in nature, in the last 40 years, a vibrant and growing set of theoretical models have arisen that demonstrate a variety of evolutionary pathways to cooperation. These pathways are not mutually exclusive solutions to the dilemma of cooperation, and different pathways may cross in ways that either promote more cooperation or debilitate it by creating conflicts. Any particular organism may have evolved to make use of one, two, all of them, or none of them. Some of the pathways or "classes of models" that we discuss below are applicable to an enormous range of species, while others are premised on a heavy reliance on high-fidelity cultural learning and thus may be restricted to humans. The second half of this chapter provides an introduction to each of these classes of models.

The Puzzle Deepens for Our Species

The puzzle of cooperation is even more interesting and enigmatic for our species. This is because the nature of human cooperation, while similar in some ways to the patterns observed in other species, is quite different in several key respects. At a macro level, human cooperation varies substantially from that of nonhuman primates in both its scale and the nature of its variability. Though the scale of cooperation in other primates rarely exceeds two or three individuals (e.g., in grooming, coalitions, and food sharing), humans in some societies, including many hunter-gatherer societies, cooperate on scales involving hundreds or even thousands of individuals (e.g., warfare, voting, recycling, and exchange networks). The scale of human cooperation varies dramatically, from societies that are economically independent at the family level—showing little cooperation outside the extended kin circle (e.g., see Johnson and Earle

[2000] on the Machiguenga and Shoshone)—to the vast scales found in chiefdoms and modern states (Richerson and Boyd 2000). Though ecological factors are certainly part of the explanation for this variation, substantial degrees of variation in the scale of cooperation can be observed among social groups inhabiting identical environments (e.g., see Atran, Medin, et al. 1999 on the Itza, Ladinos, and Kekchi Maya; and Kelly 1985 on the Nuer and Dinka). Moreover, historical sources show that the scale of cooperation in many societies has increased by orders of magnitude in historical time (Diamond 1997), thereby indicating the presence of some nongenetic evolutionary process that has been ratcheting up the scale of cooperation (a process that has not been observed in other species).[7] Finally, although primate species typically show little variation in the behavioral domains of cooperative behavior, human social groups vary substantially in their domains of cooperation. For example, some human groups cooperate in fishing, but not in house building or warfare, whereas others cooperate in house building and warfare, but not fishing. A complete approach to the puzzle of human cooperation needs to be able to explain these patterns in a manner that links humans to the rest of the natural world but, at the same time, explains our distinctiveness (Henrich 2004a; Henrich, et al. 2003).

"Intelligence" Is Not the Answer

Because the scale of human cooperation is so much greater than that found among other mammals and particularly other primates, it is a common intuition that human cooperation must result from our "superior intelligence." This intuition is likely wrong for both theoretical and empirical reasons. First, a substantial amount of theoretical work in economics, anthropology, and biology shows that more intelligence usually leads to less cooperation, not more. Cognitive capacities for strategic thinking that include planning for the future, storing data on past interactions, and more accurately assessing potential costs and benefits do not generally lead to more cooperation, as many people think (we'll discuss this in more detail below). In a complicated world with imperfect information, the skills of deception, deceit, trickery, and manipulation, which are improved by some kinds of intelligence, are more effective at destroying cooperation than capacities for tracking past interactions and planning are for sustaining it. As is often the case, it is far easier to break something fragile than to preserve it. As with fighting terrorism, an intelligent defender needs to think of every possible avenue that any potential attacker might use, whereas an attacker only has to think of one avenue that the defender has not considered.

Second, empirically the intelligence hypothesis does not lead to the kind of cooperation that characterizes our species. As just noted above, the scales of human cooperation vary dramatically across social groups and domains (even when groups inhabit identical environments), and have changed over time. It is difficult to see how the intelligence hypothesis could explain these fundamental patterns.

Finally, can we get any insight on the relationship between intelligence and cooperation by looking comparatively at other cooperative species? Besides humans, the next best cooperators in the animal kingdom are the eusocial insects (bees, wasps, ants, etc.). These insects manage to achieve massive levels of cooperation—comparable only to that achieved by humans—with relatively few neurons per individual. There are many species that have substantially more neurons (by several orders of magnitude) than eusocial insects—including all primates—but all of these species cooperate less. There is apparently no necessary relationship in nature between intelligence (measured as brain power) and cooperation. Other than a strong intuition, there is little to support the "intelligence hypotheses."

The Evolved Social Psychology of Cooperation

Here we discuss five evolutionary theories that provide potential ultimate solutions to the dilemma of cooperation. Our objectives are to provide the reader with an intuitive understanding of how evolution can lead to cooperation in humans, and to develop an understanding of the proximate psychological mechanisms—and observable behavioral patterns—from each theory. Our five classes of evolutionary models are (1) kinship (chap. 5); (2) reciprocity (chap. 6); (3) reputation (chap. 6); (4) social norms and punishment (chaps. 7 and 8); and (5) ethnicity (chap. 9). In laying these out, we will discuss both some of the fine nuances of how they work (or fail to work) and the role played by culture in both the phylogenetic (evolutionary) and ontogenetic (developmental) emergence of these forms of cooperation. Because much of this theoretical material is based on mathematical models that are beyond the scope of this book, we hope to inspire our readers to further explore evolutionary game theory and to delve into the primary literature.

The Core Dilemma of Cooperation

There is a simple core principle that underlies nearly all solutions to the puzzle of cooperation: Cooperation can evolve under circumstances in which selection takes advantage of a stable regularity that allows cooperators to preferentially bestow their benefits on other cooperators; in other words, cooperation can evolve when cooperators tend to cooperate with other cooperators. Equation (1) succinctly expresses this condition in its most general sense (derivation in appendix A):

$$\beta b > c \tag{1}$$

Here c is the cost paid by the cooperator in order to deliver benefits, b, to another individual or group of individuals. These costs and benefits are measured in units of fitness (e.g., number of offspring). β is a statistical relationship called a regression coefficient. It measures the degree to which "being a cooperator" predicts "bestowing benefits on other cooperators." In the simplest case, it is the probability that a cooperator is bestowing benefits on another cooperator. When conditions exist such that $\beta b > c$, then natural selection can favor the spread of genes that build proximate psychologies for cooperative behavior. All solutions must be able to sustain the conditions that allow cooperators to benefit other cooperators. Interestingly, in the history of the study of the evolution of cooperation (Frank 1995, 1998), this simple equation first emerged for specific cases (e.g., kinship and reciprocity) before the more general, abstract condition was derived.

The Rise and Fall of Green-Bearded Cooperators

Once this general problem is understood, a simple solution suggests itself. Imagine a gene that causes its bearer to both have a green beard and to help only other green-bearded individuals.[8] In our equation above, β would be at its maximum value of one (green-beards deliver benefits only to other green-beards), and cooperation would spread rapidly. Soon the entire world would be green-bearded cooperators, and everyone would be merrily cooperating with everyone else. However, the statistical relationship (represented by β) between bestowing help and being a cooperator must be reliable and durable. Now imagine we are in a jolly world of green-bearded cooperators, and a mutant green-beard emerges. This fellow has the requisite green beard, but he's not a cooperator. He never helps anyone, but everyone always helps him because of his lush green beard. Consequently, the mutant, and his mutant gene, will be very successful, and will eventually drive the green-bearded cooperators to extinction. The world now consists entirely of green-bearded defectors. Thus, the trick to solving the dilemma of cooperation is not so much in producing a positive β value (i.e., in creating a way for cooperators to find other cooperators), but in maintaining a reliable, stable β value.[9] This is what we call the Core Dilemma of cooperation. Next, we discuss how genetic relatedness among cooperators meets this challenge and provides one set of pathways to a particular kind of cooperation.

Kinship

Let's begin with one of the most prevalent forms of cooperation in nature.[10] Consider a mother with a gene (or genes) that causes her to experience deep positive emotions toward her offspring, which in turn often leads to costly caring for these offspring. Why would such a gene spread? Following the general logic outlined above, it could spread only if the recipient of the "help" was also a "helper"—that is, if the offspring is also a carrier of the gene that produces the psychology for helping offspring. This leads to the question: What are the chances that a mother's offspring has the helping gene, given that the mother has the helping gene? The answer can be rather complicated, but if the organism has a genetic system like humans (diploid), then the chance should be at least 50 percent. The main reason is simple. The mother and her offspring will, on average, share 50 percent of their genes in common, which means that there is a 50 percent chance that the offspring has a copy of the mother's "help your offspring" gene. Thus, the characteristic of "being one's offspring" is a reliable predictor ($\beta = 0.5$) of sharing the relevant cooperation gene(s). This kind of evolutionary process is called "kin selection" because mother and offspring (or any blood relatives) share copies of the same gene by descent from a recent common ancestor. In a sense, by causing a mother to help her offspring, the gene is probabilistically helping a copy of itself to survive and reproduce.

To see the crux of the dilemma of cooperation, consider a mutant variant of the "helper" gene (described above) that causes its bearer to feel equally affective toward all infants and juveniles, not just her own offspring. A mom bearing this genetic variant directs help and caring toward whichever infant or juvenile is most in need. Could this mutant gene spread? No, because bearers of the original variant—"help your own offspring"—would receive help not only from its own mother but from other women carrying the mutant gene. Meanwhile, the offspring of mutant mothers would receive only minimal help from their own mom (who would be spreading her help around widely), and no help from women carrying the "help your own offspring" gene. All else being equal, this would cause the "help your own offspring" gene to outcompete the "help everyone's offspring" gene.

Returning to the core principle, natural selection will favor the evolution of psychological mechanisms that allow cooperators to focus their benefits on other individuals who are likely to also be cooperators. From this perspective, our kin psychology, which is our innate preference to favor relatives over nonrelatives, and close relatives over more distant ones, represents a class of different proximate mechanisms that take advantage of the fact that some individuals in any population will tend to have the same genes by descent from a recent common ancestor (such as a parent). This fact of biology creates numerous evolutionary opportunities for natural selection to find reliable statistical regularities to exploit. "Blood relatives" have many characteristics in common in addition to the all-important cooperation gene(s), and natural selection can take advantage of these in building psychologies that can preferentially direct benefits at other individuals likely to have cooperation genes. For example, close kin, such as siblings, may have a similar appearance or smell that natural selection may use as cues to build a psychology such as "help those who look and smell like you." These similarities may be related to sharing some of the same genes (ones not related to cooperation), or they may be related to having been reared by the same female in the same nest. Even proximity can provide an evolutionary opportunity to construct a psychological mechanism capable of maintaining stable cooperation (e.g., "help those who tend to hang around the female who feeds you"). It is important to realize that natural selection does not care how a particular cue arises (i.e., it does not care about kinship per se), only that the cue is a stable and reliable predictor of who is likely to be a cooperator. In a variety of species, many aspects of reproduction create reliable statistical patterns on which natural selection can build. However, the details always depend on the specifics of the particular species. Natural selection might, for example, lead to a system that causes newborns to smell those around them during the first few days after birth, and subsequently direct benefits toward those individuals for life. This species must have a social structure that reliably places newborns among close kin, a physiology such that either individuals or close kin produce distinct (distinguishable) scents, and some way to exclude nonkin from sneaking into the nest during those first few days or faking the scent of other family members. The "exclusion" need not be foolproof for some cooperation to evolve, but the better it is, the greater the degree of possible cooperation.

In studying how kinship can solve the dilemma of cooperation, W. D. Hamilton (1964) derived the simple rule that now bears his name (Hamilton's Rule):

$$rb > c \qquad\qquad (2)$$

The reader should note the similarity between equations (1) and (2). As in (1), b is the benefits bestowed and c is the cost to the bestower. Now, however, r replaces β and represents the "coefficient of relatedness," which specifies both the average proportion of genes shared by the two individuals as a consequence of recent common descent and, more importantly, the probability that the receiver of help shares a specific gene (i.e., the cooperative gene). The parameter r is a special case of β that occurs when the foundation of cooperation is built on kinship and an organism can, by some means, direct benefits preferentially toward its kin.[11]

If we assume that by using a variety of these cues, humans (and other animals) can assess their degree of relatedness, then we should be able to predict who they will be most likely to help. This suggests that individuals should cooperate with relatives according to their "coefficient of relatedness" (r), which is the probability that they share the same gene by descent. In our species, siblings and parent-offspring have the highest r at 0.50 (except for identical twins, who have $r = 1$). Assuming only one line of descent (no interbreeding), grandparents and their grandchildren and half-siblings have the next highest relatedness at $r = 0.25$. First cousins are related at $r = 0.125$. Using the evolutionary logic of kinship, parents and their children, and siblings should cooperate a fair amount. First cousins may cooperate a bit, but more distant relatives have too little relatedness to cooperate with one another. Among siblings, for example, the fitness cost (in terms of survival and reproduction) must be less than half of the fitness benefits delivered before cooperation would be favored. To get an intuitive sense of this, consider to whom you would give a kidney. In the United States, 86 percent of kidney donations come from close kin, while less than one-half of one percent come from anonymous strangers (and thousands of people die every year waiting for kidneys).

Culture and Kinship

Of our five evolutionary avenues to cooperation, cultural evolution has probably had the least influence on our kin psychology. Nevertheless, culture still influences kinship in several important ways. First, cultural evolution has created a variety of kinship systems, which provide categories, labels, and norms that influence the operation of our basic kin psychology. All human languages have kin terms that systematically specify categories of kin, types of relations, and degrees of closeness. Although these systems often reflect genetic related-ness, they also diverge from strict relatedness and reclassify people in interesting ways. Among cues that our psychology likely uses to assess relatedness are culturally evolved categories and kin labels. Cultural systems that define kinship can influence behavior in powerful ways. For example, in our own society, adopted children are treated indistinguishably from genetic children (Daly and Wilson 1999)—although stepchildren are substantially more likely to be the victims of violence and homicide at the hands of their unrelated parents.

A particularly interesting case of how culturally learned beliefs influence the operation of our kin psychology involves what is known as "partible paternity." In New Guinea and throughout lowland South America, cultural beliefs hold that children are formed by repeated ejaculations of sperm into a woman. In fact, many of these cultures maintain that a single ejaculation cannot create a viable pregnancy, and men must work hard with repeated ejaculations over many months to create and maintain a viable fetus. Lacking rigid restrictions on extramarital sex, pregnant women will seek out "additional fathers" for their future child. Or, in some of these societies, periodic rituals prescribe extramari-tal sex and thereby formalize the creation of multiple fathers. These secondary fathers—often named at birth by the mother—are expected to contribute to the welfare (in the form of meat and fish) of their children, although not as much as the primary father (the mother's husband). Interestingly, detailed studies show that children with exactly two fathers are more likely to survive past age 15 than children with either one father or three or more fathers. By introducing cultural beliefs about how reproduction works, cultural evolution has modified how kinship is assigned, which in turn has influenced survival (Beckerman and Valentine 2002).

Culture also influences the operation of our kin psychology by creating norms that affect who lives with whom. In turn, these residence patterns can influence the availability and strength of certain kinship cues. For example, in patrilocal societies, married couples take up residence in or near the husband's father's household. This means that children will be differentially exposed to their father's family, perhaps leading them to "feel more kinship" toward their father's mother, father's father, father's siblings, and father's brother's children than toward their genetically equidistant relatives on their mother's side. This can occur if children are using cues of proximity, phenotypic similarity, and/or certain kinds of scent, especially if these cues are particularly operative during some sensitive developmental window.[12] Along the same lines, in some societies where men and adolescent boys live separately from sisters and mothers, young children will see substantially less of their older brothers and fathers than in societies in which they cohabitate. Often in these societies, the mother's brother takes the role most readers would traditionally assign to the father.[13]

We continue this discussion in chapter 5 by bringing empirical findings to bear on this theoretical work in four ways. First, we review some of the laboratory work showing that humans use a variety of cues to figure out who their kin are, and how much to help, cooperate, and/or trust. Second, using data from our ethnographic study of Chaldeans in metro Detroit, we confirm a robust finding from small-scale societies (Chagnon and Irons 1979; Cronk, Chagnon, and Irons 2000) that people restrict their most costly forms of cooperation to close kin. Third, we show that although Chaldeans have an explicit cultural ideology that all Chaldeans are "one big family" and that everyone is related, no one is fooled behaviorally; the lines of costly cooperation are clearly drawn at the outskirts of the immediate family. This addresses the claim that large-scale human cooperation in contemporary societies results from the faulty (maladaptive) operation of a psychology designed for cooperating in the small-scale, kin-based societies that many believe characterized human ancestral life. Fourth, we discuss how kinship and cultural expectations about behavior brought from Iraq combine to allow many Chaldean storeowners to outcompete non-Chaldeans economically. As part of this, we suggest that Chaldean cultural beliefs are currently shifting toward models typical of other middle-class Americans. This case material indicates that although kin psychology is constant, adaptive cultural learning processes may be gradually shifting cultural beliefs, which is altering the details of who helps whom. This shift has important economic, social, and educational effects on Chaldeans.

Reciprocity

Reciprocity represents another well-studied class of potential solutions to the evolutionary puzzle of cooperation (Trivers 1971). Individuals, by applying tit-for-tat exchange strategies, can preferentially associate themselves with other cooperators, and thereby increase both their chances of bestowing their benefits on other individuals with the same genes and increase their likelihood of receiving benefits from others by putting themselves in the company of other reciprocators.

Progress on reciprocity as the answer to the dilemma of cooperation has differed from kinship in two important (and often unrecognized) ways. First, despite literally hundreds of papers on the topic (Axelrod and D'Ambrosio 1994 lists 209 publications from 1987 to 1993), the theoretical conclusions derived for the mathematical models and computer simulations are substantially more ambiguous, nuanced, and qualified than those for kinship. Second, the empirical evidence for reciprocity-based cooperation in nonhuman species is scant (Hammerstein 2003), especially when compared to the evidence for kin-based cooperation. Nevertheless, in our species, reciprocity consistently emerges in both ethnographic and experimental studies.

Below, we deal with these two aspects of reciprocity by synthesizing the qualitative findings from a substantial body of theoretical work and linking these to our broader empirical efforts. Unlike most general treatments, we will set this research within the framework of the Core Dilemma, and integrate it with our species' heavy reliance on cultural learning. This last connection allows us to explain why reciprocity-based solutions are rare in the rest of nature but plentiful and diverse in human societies. Theoretically, there are two take-home messages here: (1) reciprocity-based solutions to cooperation are much less robust than many scholars think, and (2) these solutions are so prominent in humans because of (not despite) our evolved cultural learning capacities. Empirically, we use the qualitative insights developed below to generate a series of predictions about how our evolved reciprocity psychology works. In chapter 6, we substantiate some key predictions using laboratory data and then show how reciprocity illuminates some important aspects of Chaldean social life.

Our theoretical framework shows that the evolution of cooperation requires that benefits be preferentially bestowed on cooperators. Above, we explained how kinship can enable this by providing reliable cues that help identify the cooperators. Reciprocity-based solutions, however, use different mechanisms to support cooperation among nonkin. In this class of solutions, natural selection favors individuals who can use the past behavior of other individuals as an indicator of whether they are a cooperator (or reciprocator) or not. If a person's past behavior suggests that she may be a reciprocator (reciprocates cooperation), then natural selection can favor a psychology that promotes bestowing benefits on that person. There are two reasons for this selection pressure. First, an individual's past history may act as a cue that she carries a "cooperative/ reciprocator gene." By bestowing benefits on those with a history of cooperation, cooperators may be able to preferentially direct benefits toward others with the same gene. Second, and more important, bestowing benefits on reciprocators can cause a return flow of benefits, in some fashion, back to the bestower. In this case, the specific genes carried by the other individual are irrelevant. What matters is that bestowing benefits causes a return flow of benefits back to the one with the "reciprocation gene." There are two sources of information about past behavior that might help diagnose whom one should bestow benefits upon. These sources lead to the two commonly discussed forms of reciprocity: (1) direct reciprocity, based on direct, personal experience with partners, and (2) indirect reciprocity, which involves getting information about potential interactants by observing them with others, or by gathering reputational information (culturally transmitted information) about their past behavior with other individuals.

Direct Reciprocity

The intuition behind direct reciprocity is simple: if you help me, I will help you. If you stop helping me, I will stop helping you. Direct reciprocity depends on direct, ongoing experience between interacting individuals. To illustrate, suppose Joe and Natalie are each small-business owners. Each business goes through an annual period of a cash shortage at different times of the year. If each business can get through this period, each can make an overall profit, but if it cannot, the business makes no profit. Getting through the period of scarcity requires $100. If Joe and Natalie give each other money when the other needs it, both make $300. If neither gives money to the other, both make zero profit. If Natalie gives Joe money during his scarce period, but Joe decides not to give

Table 3.1 Payoffs to Decisions in Each Year

	Cooperate (give money)	Defect (don't give money)
Cooperate	300, 300	−100, 400
Defect	400, −100	0, 0

Natalie any money during her scarce period, Joe will make $400 ($300 plus the $100 he did not give out) and Natalie will lose $100 (zero profit minus $100 given to Joe). Thus, by "defecting" (not giving the money), Joe does relatively better than Natalie (by $500) in that year, and Natalie may go out of business, leaving Joe's business practices to proliferate as people see his financial success and copy his strategy. Over time, as more and more people act like Joe, no one will give any money, and everyone will go out of business. If Joe and Natalie had sustained cooperation, then the pair would have done much better overall than they did when Joe defected. Had they sustained cooperation, after 10 years each would have a $3,000 profit. Instead, Joe made $500 the year he defected but never made any more in subsequent years because his trading partner, Natalie, went out of business and lost $100. Over 10 years, their combined profit was only $400.

To study the kinds of strategies or behavioral rules that will succeed in maintaining cooperation through reciprocity, theorists have formalized the above situations in an abstract format called the Prisoner's Dilemma (PD). This mathematical formalization allows researchers to systematically study the properties of various strategies, and other variables, in solving the dilemma of cooperation. In the PD, individuals are paired at random, interact repeatedly for some number of rounds (like Joe and Natalie giving money to each other each year), and receive payoffs based on their decisions and those of their partner. Individuals using strategies that give them higher payoffs relative to other strategies will increase in relative frequency compared to those using strategies that yield lower payoffs. The game matrix (table 3.1) illustrates the payoffs and captures Joe and Natalie's business situation.

Some of the earliest work by Robert Axelrod and W. D. Hamilton on direct reciprocity using this approach suggested that simple reciprocating strategies could generate long-term, stable cooperation (Axelrod and Hamilton 1981). One of the central analytical findings in Axelrod's book (1984), *The Evolution of Cooperation*, is that repeated interaction with the same player is merely another way to address the Core Dilemma in the evolution of cooperation (also see Maynard Smith 1982). If we take a simple reciprocating strategy that cooperates in the first round and in subsequent rounds does whatever the other guy just did (tit-for-tat) and assume this strategy is common in a population, the condition for the reciprocating strategy to remain common against low frequency invading defectors is:

$$\omega b > c \qquad\qquad (3)$$

Look familiar? Here, as above, b and c are the costs and benefits of cooperation, but now ω (replacing β from [1]) is the probability that the interaction with the same individual will continue in the next round. Essentially, ω is a measure of how long the cooperation can continue if both individuals keep cooperating. In this way, prior interactions with the same person in a repeated sequence provide a cue about whom to preferentially interact with to receive a flow of benefits. The longer the interaction continues with the same reciprocator, the greater the amount of cooperation that can be sustained, because the individuals receive more total benefits.

Although several of the general findings put forth by Axelrod (1984) have held up well (we discuss those shortly), other heralded findings have not stood the test of time and have been overinterpreted. Perhaps the starkest example of this is the reciprocating strategy "tit-for-tat" (TFT), which was thought to be the most robust and simplest solution to the problem of cooperation among nonkin. All the hoopla surrounding TFT resulted in a number of misunderstandings and led a well-known evolutionary biologist to write in a textbook that TFT "is superior to all other [strategies] in playing repeated games of prisoner's dilemma" (Trivers 1985: 391). Not only was such a statement not supported by the existing work at the time, but soon after 1985, a long parade of papers showed that TFT wasn't even "pretty good" in many situations.[14] For example, when information about the payoffs or behavior of one's partner is less than perfect, or when one is in a large group, TFT does not lead to very much cooperation, and much better strategies exist. In these and other situations, there are many ways to preserve cooperation, but this diversity itself creates an evolutionary challenge that culture can robustly solve.

Overall, the important intuitive understandings about reciprocity that have arisen since the mid-1980s suggest that (1) reciprocity-based solutions are substantially more complicated than many thought, and (2) although these complicated, contingent patterns lack the desired simple elegance sought by many, they appear to be consistent with empirical patterns among human societies and across species. Below, we synthesize what is known about reciprocity in an effort to paint a coherent image of the evolution of cooperation via direct reciprocity. Using this image, we sketch the patterns of behavior that we expect our evolved psychology to produce.

On the positive side, although no simple strategy exists that yields anything approaching a robust reciprocity-based solution to the dilemma of cooperation, much of this theoretical work suggests that some types of reciprocating strategies can lead to substantial amounts of cooperation in a wide range of circumstances as long as group size remains small (dyads, triads, etc.) and interaction endures for a sufficiently long time (a large ω). Though some kinds of reciprocating strategies can be successful in particular circumstances, the details of such strategies vary tremendously. To give a sense of which aspects of the cooperative environment influence the success of different kinds of reciprocating strategies, we review five key factors: (1) group size, (2) noise, (3) variation in the duration of interactions across partners, (4) the ecology of other strategies, and (5) social network and partner choice.

Group Size

The seeming success of TFT and other reciprocity-based strategies in interactions between two individuals in much of the early work led many to assume that these dyadic findings would extend to explain cooperation among larger groups (public goods problems or "*n*-person cooperative dilemmas"). Rather than two individuals helping one another, the *n* players in public goods games contribute to the benefit of everyone in the group. If one assumes that individuals occasionally make mistakes (defecting when they meant to cooperate), then reciprocity-based strategies (such as TFT) do not generate cooperation in larger groups to nearly the degree that they do in dyadic situations.[15] In fact, the capacity for reciprocity to maintain cooperation decreases geometrically as the group size increases. At the sociological level, this line of theoretical work suggests that we should not expect direct reciprocity to be the primary factor in maintaining cooperation in large groups. That is, direct reciprocity cannot be used to explain cooperation in such things as warfare, voting, recycling, or food sharing in small-scale societies. At the psychological level, this finding predicts that human psychology should be motivated by direct reciprocity *only* when the cooperative unit is small—dyads should be the preferred group for direct reciprocity. From the perspective of institutional design, cultural evolution should exploit our evolved reciprocal psychology by creating institutions that—among other things—partition cooperation into small, enduring social groups.

Noise

Reciprocity-based strategies of all kinds are entirely reliant on the information that individuals receive about the past behavior of their partners. Given this, what if individuals receive inaccurate information about what their partners did or about the payoffs they received in previous interactions? This is certainly a potential problem in the real world, as information is often ambiguous or lacking.[16] What if Natalie figured that Joe had decided to withhold the money when in fact Joe's check got lost in the mail? Joe tried to cooperate, but Natalie perceived him as a defector. Or what if your partner helps you move furniture for the full 6 hours that you wanted him, but because of some confusion, you thought he was there for only 4 hours (and you helped him 6 hours last time)? What do you do when he asks you for 6 hours of help next time? Do you stop reciprocating entirely (perhaps you think he's taking advantage of you), do you reciprocate only 4 hours (playing TFT), or do you help him for a full 6 hours?

Error or noise in this kind of information creates a trade-off between a strategy's PROVOCATIVENESS and a strategy's GENEROSITY. An individual using a PROVOKABLE strategy shows a hair-trigger willingness to stop cooperating as soon as he has any indication that his partner is not fully cooperating. In contrast, a person with a GENEROUS strategy reciprocates with more than she is given; if you give me 4 hours of your time (or at least I think that you did), I will give you 6 next time. TFT is PROVOKABLE but not GENEROUS at all, because a person using TFT stops cooperating as soon as it appears her partner has defected. Not being GENEROUS in a noisy environment can lead to substantially reduced amounts of cooperation, as misunderstandings lead to irredeemable losses in cooperation. To see this, suppose two people using TFT strategies were interacting—let's call them TFT[1] and TFT[2]. On round one TFT[1] gives 10 (the maximum) to TFT[2], but because of the noisy environment TFT[2] thought it was only 8. Then, in the next round TFT[2] returns 8 to TFT[1], but TFT[1] thought he received only 7, and he replies with only 7. Now, although some misunderstandings can drive exchanges back up, the average exchange would not be near 10—actually this setup leads to an average exchange of 5 in the long run. In contrast, a GENEROUS strategy can deal with this. Suppose two people with TFT+2 strategies (TFT+2 gives back two more than it was given) are playing in a noisy environment. If TFT+2[1] gives 10 to TFT+2[2], but TFT+2[2] believes that only 8 were sent, she will still give 10 back to TFT+2[1]. Analysis shows that GENEROSITY leads to more cooperation in noisy environments than strict-return strategies like TFT. However, the effectiveness of being GENEROUS depends critically on what other strategies are lurking in the population, because GENEROUS strategies that can maintain effective cooperation in noisy environments are susceptible to exploitation by crafty strategies (e.g., TFT-1 gives 1 less than she believes she was given) that hide under the cover of noise and exploit GENEROUS strategies. If you are in a nice environment, one in which most people cooperate in the first round of an interaction and then use a reciprocating strategy, GENEROSITY makes the most of cooperation. But if exploiter strategies emerge in sufficient numbers, being highly PROVOKABLE is the only defense. Interestingly, GENEROUS strategies can thrive if there are a sufficient number of PROVOKABLE strategies to keep the crafty defecting strategies at bay. And, GENEROUS strategies bring out the best in PROVOKABLE strategies such as TFT. In noisy environments, PROVOKABLE strategies like TFT cooperate at much higher levels with GENEROUS strategies like TFT+2 than with other PROVOKABLE strategies, while keeping lurking exploiter strategies in check.[17] Strategic diversification can help immunize a population against invading defectors and thereby promote cooperation by reducing the inherent trade-offs between being PROVOKABLE versus GENEROUS.

Besides favoring generosity, noisy environments can also favor CONTRITION. People using contrite or remorseful strategies accept punishment (in the form of a defection) after they have defected. A person with a CTFT (contrite TFT) strategy cooperates in the first round and then reciprocates cooperation and defection in subsequent rounds. But if CTFT's partner defects after a mistaken defection by CTFT, then CTFT will cooperate in the next round despite the partner's defection. After this, CTFT will return to playing TFT. This allows CTFT to repair cooperative relationships that would otherwise be destroyed by some kinds of errors (Boyd 1989). For example, if CTFT is interacting with TFT and CTFT accidentally defects when he meant to cooperate, he will "accept" TFT's defection on the next interaction and cooperate one time after this. This cooperation will bring out cooperation from TFT, and the pair will return to cooperative interaction. Had two TFTs been interacting, all cooperation would have ceased.

CONTRITE strategies are, however, susceptible to "errors in perception" (Boerlijst, Nowak, and Sigmund 1997). With these errors, a mismatch arises between what the individuals in the interaction believe happened. For example, Natalie may believe Joe defected, when in fact Joe cooperated and believes he cooperated. Even if Joe uses a contrite strategy, the pair won't be able to repair their relationship because Natalie will defect in the next round, and Joe won't understand why (because he thinks he cooperated). CONTRITION cannot always save you from the destructive effects of errors.

One final aspect of noisy environments to consider is a memory for past interactions. While some kinds of memory can be an effective means to sustain cooperation in noisy environments, longer memories are not necessarily better and can actually lead to less cooperation. Longer memories can lead individuals to adapt themselves to the *noise*, rather than the opportunities for cooperation available from the array of other strategies in their environment. Additionally, more memory is cognitively costly, especially as a person's number of partners increases (Bendor 1993, 1987). The importance of memory in sustaining cooperation via reciprocity is often exaggerated. It has been frequently assumed that humans cooperate more than other animals because they can remember a longer history of interactions. This is a case in which common intuitions fail in the face of rigorous formal models.

Variation in Duration of Interactions across Partners

Under a wide range of conditions, reciprocating strategies that lead to long-term cooperation are generally NICE. A strategy is considered NICE if it cooperates in the first round of an interaction and it is these strategies that are most likely to be successful at achieving and maintaining cooperation. Strategies are not always NICE or non-NICE. Rather, whether or not a strategy is NICE should depend on cues about how long a specific set of interactions might go on, and whether there are other reciprocating strategies out there to cooperate with—if everyone is a defector, being NICE does nothing for cooperation. One implication of this is that if a population has individuals who vary in how long they will stay around (different ω's), and this difference cannot be easily detected, then people with non-NICE strategies who wait until they are sure they are with a long-term interactant before initiating cooperation can be the most successful. By waiting, non-NICE strategies can effectively target their cooperation at other long-term interactants. This suggests that natural selection should favor individuals who are (1) NICE to long-term interactants, (2) non-NICE and non-cooperative to short-term interactants when such individuals can be distinguished from long-termers, and (3) initially wary (non-NICE) but eventually cooperative when the population is a mix of long-term and short-term interactants, and they cannot tell the difference (Boyd 1992).

The Ecology of Other Strategies

Above we have hinted that the success of a strategy depends on the other strategies that exist in the population. This turns out to be a remarkably general property of reciprocity-based cooperation. For any strategy one can devise, there is a combination of other strategies that will destroy it! In a variety of situations, it has even been possible to prove mathematically that no reciprocating strategy is safe and robust against other combinations of strategies (Bendor 1993; Bendor and Swistak 1997; Boyd and Lorderbaum 1987; Farrell and Ware 1989; Lorberbaum 1994; Lorberbaum et al. 2002). This means that successful individuals need to be able to rapidly adapt their strategies to the changing balance of other strategies in the population. This presents a particularly prickly problem for natural selection to solve, since shifts in the social-ecological balance may occur rapidly, even within the lifetime of an individual. Because there are so many possible social ecologies, individuals in many species (especially humans) are likely to encounter ecologies that their species has never experienced in the evolutionary past. As we'll explain, we believe that the human adaptation for cultural learning provides one of the few means through which natural selection can meet this—the real challenge—of the evolution of cooperation via direct reciprocity.

Social Network and Partner Choice

The models described above are complemented by recent simulation work that allows individuals with a limited memory of others' past behavior to selectively associate with preferred partners, thereby spontaneously forming social networks of reciprocators (Hruschka and Henrich 2006). This work nuances past work in a variety of ways, but most important for our purposes is that successful strategies (1) weight the most recent past interactions more than TFT and (2) are only NICE when lacking preferred partners. This means that partners weigh the history of their relationship together in deciding whether to cooperate, defect, or discontinue the interaction and search for another partner. Such work also robustly shows that once individuals have established a limited set of preferred partners, these successful strategies defect upon encountering nonpartners from their group. That is, being NICE should depend on whether one has established a reliable set of reciprocators. NICENESS is conditional.

Why Reciprocity-Based Cooperation Is Rare in Nonhumans

The effect of small shifts in the ecology of other strategies in the environment and the importance of variables such as noise and group size have been underappreciated by many evolutionary scholars. We think that the right take-home message from all this theoretical work is that the genetic evolution of cooperation via direct reciprocity is not a particularly robust solution for most animals in complex exchange contexts. In a sense, the theoretical work shows that there is no way to build an "all-purpose reciprocity machine." Even if one could somehow encode the rules for several different reciprocity strategies into the genes of an organism, there is a virtually infinite number of strategies one would need. This creates a combinatorial explosion of potential contingencies that would have to be built into the genes.[18] The situation is especially acute in humans, given the variety of cultural and physical environments in which we operate. Consistent with the above logic, the empirical record shows little evidence for reciprocity in nonhumans (Hammerstein 2003), and the nonhuman examples that do exist are relegated to special-case situations.[19]

Enter Cultural Transmission

Though this line of theoretical argument is consistent with nonhuman data, we are left with a puzzle in the human case. In contrast to nonhumans, the available ethnographic and empirical evidence shows that direct reciprocity is a strong and recurrent pattern among our species and across cultures. Yet, interestingly, the details and circumstances of direct reciprocity are highly variable across contexts and human social groups (Fiske 1991). We argue that the combinatorial explosion is solved in humans through cultural transmission. The adaptive nature of cultural learning works with a psychology for reciprocity that coevolved with culture to create "custom-fit" strategies (solutions) that are dynamic and adaptable, to meet the challenges created by the ever-shifting mix of strategies described above.

To understand how culture solves the combinatorial explosion of reciprocity, it is important to recognize that (1) cultural evolution occurs at much faster rates than genetic evolution (e.g., novel practices can spread through a population in a single generation; see E. Rogers 1995); (2) cultural learning processes allow populations to rapidly adapt to novel situations without any genetic change;[20] and (3) cultural transmission can, to a degree, construct its own environment (e.g., written records reduce the "noise" in payoffs between interactants, or institutional structures can turn public goods problems into dyadic interactions; see chap. 6). Genetic evolution will continually introduce novel mutations (that lead to new strategies), but cultural evolution will be able to counter such mutations by building new culturally evolved strategies or altering the social environment by building new forms of organization (e.g., communal housing, ID cards, police forces, neighborhood watch), at a rate much faster than genetic evolution. Cultural learning processes, like prestige-biased transmission, will allow individuals to adapt their reciprocity-based strategies to shifting social and physical ecologies within their lifetime—we've already observed this in the laboratory studies described in chapter 2. The effect of cultural evolution on reciprocity-based cooperation accounts for the empirical fact that humans in different societies use reciprocity to differing degrees and in different contexts. Our approach can account for both the presence of local variation and the universal aspects of human reciprocity.

Though genetic evolution alone cannot provide a solution to the real challenge of direct reciprocity, it can promote an individual's success by providing the learning mechanisms, biases, and default settings that allow one to rapidly and effectively acquire locally useful strategies. We suspect that our reciprocity psychology is a product of the coevolution between genes and culture, and rather than supplying rules like TFT, our psychology provides the machinery for rapidly learning the locally successful rules of reciprocity from other people and from experience. By imitating successful individuals, for example, people can acquire the locally adaptive strategic nuances that fit the local ecology of strategies, the noise, and so on. Projecting back into time, we might imagine that once our capacities for cultural learning evolved genetically, then cultural evolution produced the first simple rules of reciprocity (via imitation[21]), in the same fashion that cultural evolution produced early stone tool traditions. Following on the heels of cultural evolution, genetic evolution via natural selection may have favored "learning genes" that allow naïve individuals to more rapidly acquire reciprocity-based rules (strategies). With these new learning biases in place, perhaps cultural evolution was able to solve an increasing number of cooperative problems using a variety of different reciprocity-based strategies in different contexts and places. The outcome of this interactional process might be an evolved reciprocity psychology that operates through (at least partially), and is dependent on, cultural learning to acquire the adaptive strategies for social behavior. This evolved psychology should assist individuals in acquiring effective strategies, avoiding exploitation, and identifying good long-term partners. In a sense, genetic and cultural evolution seem to have teamed up to solve a variety of cooperative dilemmas through the logic of direct reciprocity.

Patterns of Direct Reciprocity

Based on theoretical intuitions built up by reviewing the findings of numerous models of the evolution of reciprocity, we now have an idea (a complex of hypotheses) about the general shape of this cognitive and emotional adaptation. Because our reciprocity psychology coevolved with culture, some of what's below can be understood as (1) mechanisms (designed by natural selection) that assist imitators in inferring a model's underlying strategy for success based on the model's overt or public behavior, or as (2) calibrations based on direct evaluations of the local ecology, involving such things as the noise in payoffs, the presence of other strategies, and the future time horizon. To the first, inferential mechanisms are an important part of cultural learning as learners want to copy underlying mental representations (strategies, emotions, etc.) but can obtain only partial information by observing behaviors (which include, but are not limited to, linguistic acts). Such inferential mechanisms or context biases (see chap. 2) organize observed behavioral information, constrain inferences, supply default settings, and direct attention to certain aspects or patterns of behavior. This is important and possible because, as we saw above, although no specific strategy is robust, successful strategies have several characteristics that can be calibrated to fit a wide range of environments. In combination with other cultural learning mechanisms, including prestige-biased transmission and conformist transmission (chap. 2), these inferential mechanisms allow individuals to adapt and calibrate their strategies to local situations, including the dynamic ecology of other strategies. For the second, our psychology should take advantage of direct information about the relevant aspect of successful reciprocity whenever such information makes itself available. Drawing from the above, our reciprocity psychology should produce some observable characteristics, such as:

1. NICENESS. When should a successful strategy cooperate on a first interaction and how much cooperation is appropriate? NICENESS is essential to creating cooperation in many environments and may be a default strategy, especially within local groups (or ethnic groups).

    a. However, when groups are large ($n > 5$ to 10), the expected length of the interaction is short or highly variable, and/or the ratio of benefits to costs (b/c) is small, individuals should tend to adopt SUSPICIOUS strategies (i.e., strategies that aren't NICE in the first round, but can be persuaded to cooperate by some very NICE strategies).

    b. Once successful strategies have established a small set of preferred partners, they will defect on nonpartners in noisy exchanges—conditional NICENESS. That is, NICENESS is readily used with a subset of people but not generally extended to all members of a local group.

    c. Nevertheless, some NICENESS can be sustained among nonpreferred partners in tightly controlled culturally prescribed exchanges that involve little noise (for example, with in-kind 1-to-1 exchanges). In chapter 6, we will discuss how cultural evolution promotes NICENESS (and deals with the constraints of our reciprocity psychology), and thus promotes sustained cooperation, by (a) turning an n-person cooperation dilemma into a dyadic situation, (b) guaranteeing that individuals will stick around for the long run, and (c) using kinship to transcend the life of an individual and extend the time horizon into the future (using kinship to make ω large, a time horizon can extend over generations). We will also explain how culture can solve other aspects of NICENESS by setting standards for how NICE to be on a first interaction—without culture this is completely unspecified, and there are no standards or expectations.

2. GENEROSITY/CONTRITION versus PROVOCATION. How GENEROUS, CONTRITE, and PROVOKABLE should a strategy be in the local environment and ecology? GENEROSITY should be favored in environments (contexts) and ecologies that are both (a) noisy and (b) mostly reciprocal (with a significant proportion of reciprocating strategies and few defectors). When little noise clouds the payments of one's partners, or when defectors appear plentiful, individuals should prefer strategies that are more PROVOKABLE.

3. MEMORY. How much of a partner's past behavior does one remember and incorporate into a decision about reciprocation? Overall, individuals in noisy environments should use selective forms of longer-term memory (e.g., remember only big defections) for only a small cadre of valued partners. For a larger set of partners, memory capacities will be stressed so individuals should prefer strategies that require little memory, but then

restrict their interactions to low-noise environments and deploy more PROVOKABLE strategies. As just noted, in chapter 6 we will discuss how culture can establish different domains of cooperation in the same social group. In those domains that involve a large number of potential dyadic interactions and a means of reducing the noise in payoffs, cultural evolution favors fairly strict 1-for-1 in-kind exchanges, which make interactions easily trackable and less noisy, thereby allowing short-term accounting for past behavior. In other domains, such as those involving a small number of individuals who make a wide range of different kinds of exchanges (creating lots of noise), favored strategies will involve GENEROSITY, CONTRITION, CONDITIONAL NICENESS, and some kind of longer-term accounting.

Table 3.2 summarizes the patterns described above. Although it seems more likely that the different parameters of reciprocity strategies (e.g., NICENESS) are more accurately thought of as continuous, Table 3.2's trichotomy provides a condensed, heuristic presentation of the ideas just discussed.

Indirect Reciprocity and Reputation

As mentioned above, reciprocity-based approaches to cooperation can take two forms: direct and indirect reciprocity. In direct reciprocity, favors are exchanged directly and repeatedly between individuals: A helps B, and, in return, B helps A. Under indirect reciprocity, individuals interact with each other only occasionally (sometimes only once), but now—before interacting—individuals receive information about the past behavior of the individual with whom they are about to interact. For example, A and B interact, and A defects on B (who cooperates); then A and C meet, but because C (who usually is a cooperator) knows about A's past behavior with B, C defects instead of cooperating. If C had met B instead of A, C would have cooperated because C would know that B

Table 3.2 Heuristic Categories of Direct Reciprocity

Categories of partners	Context and ecology	Psychology and behavior
	Substantial noise—exchanges across many domains	CONDITIONALLY NICE GENEROUS CONTRITE
Close friends	High b/c Longer memories of important interactions Small # of preferred partners (memory constraints)	
Distant friends, and other acquaintances	Low noise—in-kind, 1-for-1 exchanges Medium b/c Short memories of interactions Potentially large # individuals	LIMITED NICE PROVOKABLE NOT GENEROUS
Others	n-person dilemma (public goods situation) Short time horizon (low ω) Low b/c	SUSPICIOUS PROVOKABLE

cooperated. C's knowledge of the behavior of A or B from previous interactions corresponds to what we commonly refer to as "reputation." The incorporation of reputation either by directly observing another's behavior or by receiving information from others (presumably connected to the experience of others) about past behavior provides a possible mechanism to explain cases of cooperation that involve neither kin nor long-term interactions (Leimar and Hammerstein 2001; Sugden 1986; Wedekind and Milinski 2000). As we will discuss in the next section and in chapter 7, the importance of reputation and the practicalities of disseminating reputational information creates a close relationship between two different pathways to cooperation: indirect reciprocity and social norms. In fact, social norms lay a foundation that dramatically strengthens the power of indirect reciprocity.

Indirect Reciprocity: Solving the Core Dilemma, Again

It should be clear by now that cooperation can evolve according to the degree by which individuals can preferentially bestow their benefits on cooperators. Theoretical work on indirect reciprocity reveals the Core Dilemma once again. This research shows that reciprocating strategies can be maintained against invading defectors when (Nowak and Sigmund 1998a; Nowak and Sigmund 1998b):[22]

$$\phi b > c$$

As above, b is the help contributed and c is the cost of helping to the helper, but now ϕ (replacing β in (1)) is the probability that the helper accurately knows the reputation of the person to be helped.

In approaching indirect reciprocity, one must realize that there is significantly less theoretical work than we have for direct reciprocity.[23] However, consistent with the simple equation above, theoretical and computer simulation research shows that the availability of accurate reputational information is the key to indirect reciprocity's ability to solve the puzzle of cooperation (Brandt and Sigmund 2004; Leimar and Hammerstein 2001; Ohtsuki and Iwasa 2004; Panchanathan and Boyd 2003). This suggests that, all other things being equal, variables such as the size of the cooperative group (the number of individuals in any given interaction), the population size (the number of individuals in the pool of potential interactants), the density of social connections between individuals in the population, and people's beliefs about gossip will strongly influence the effectiveness of indirect reciprocity. More specifically, the larger the cooperative group, the less a cooperator will be able to direct his help preferentially to individuals with good reputations, which means that a standard indirect reciprocity model cannot sustain significant cooperation in groups much larger than dyads or triads.[24] Likewise, the larger the population, the less likely it will be that an individual will have the necessary reputational information about a particular interactant, the more bits of reputational information each individual will have to store in memory, and the less accurate that information will be. Less dense social networks imply that a cooperator is less likely to have (or be able to get) accurate information about a particular individual. Finally, culturally acquired beliefs about how much to gossip and how accurate that gossip must be can greatly influence the effectiveness of indirect reciprocity. Individuals, for example, may be punished by others for spreading inappropriate or false gossip. Such sanctioning is a kind of altruistic punishment in a public goods situation, and is quite difficult to explain without the social norms (and its supporting psychological mechanisms) discussed below. However, once this punishment of those breaking the gossip norms is in place, much greater amounts of cooperation can be explained by indirect reciprocity.

Although theoretical work on indirect reciprocity is just reaching childhood, some important progress has been made on understanding how people should calculate and use reputational information. For calculating how actions turn into reputations, three broad classes of strategies have been pursued. These strategies differ in how the actions ("help" or "not help") of others are judged. Drawing specifically on Brandt and Sigmund (2004), three reputational *assessment* strategies are:

1. SCORING. This strategy condemns anyone who fails to help another, (regardless of reputation), when given the chance. By "condemns" we mean that the assessment strategy prescribes assigning a bad reputation to that individual.
2. STANDING. This strategy condemns anyone who fails to help another *with a good reputation*.
3. JUDGING. This strategy condemns anyone who fails to help another with a good reputation and further condemns anyone who helps those with bad reputations.

Once you can calculate the reputation of people, there remains the question of how to use this calculation. This work addresses how an individual should incorporate his own reputation with that of the helpee in deciding whether or not to help. Four appealing combinations of *action* strategies are:

1. HELPEE–this strategy helps whenever the helpee has a good reputation
2. SELF–this strategy helps when the individual (the *helper*) has a bad reputation
3. AND–this strategy helps when the helpee has a good reputation and the helper has a bad reputation
4. OR–this strategy helps when either the helpee has a good reputation or the helper has a bad reputation.

Computer simulations and analytical work indicate that no single combination of *action* and *assessment* strategies (e.g., SCORING and HELPEE) dominates across all conditions; dominance depends on the local environment, as is the case with direct reciprocity. However, with regard to action strategies, HELPEE and OR out-compete SELF and AND. That is, strategies that *limit* their helping to those with good reputations tend to outcompete strategies that help those with bad reputations. So action strategies need to consider the reputation of others, not merely how to influence their own reputation. Interesting in this regard is the fact that strategies that always help (regardless of reputation—unconditional altruists) are highly destructive to cooperation.

In terms of *assessment*, STANDING strategies tend to be more prevalent than JUDGING and SCORING, but all three can remain present in a population. JUDGING is even robust in conditions with very high fidelity reputational information, but drops out quickly as noise increases. SCORING strategies, as long as STANDING strategies are present, can be maintained at a significant level in a population, although often below that of STANDING (Brandt and Sigmund 2004). Overall, this work shows that reputation should penalize individuals who fail to help individuals with good reputations but should not necessarily penalize those who help individuals with bad reputations (Ohtsuki and Iwasa 2004).

As with direct reciprocity, while theoretical models show that reputation can sustain cooperation among pairs of individuals, individuals will need some mechanism to tune or calibrate their general reputational psychology to local environments. Cultural learning can provide a means of figuring out what the standards of "good" are, whether a person who helps a "bad person" is "good" or "bad," and whether to consider one's own reputation in deciding whether to help or not help someone with a bad reputation. In chapter 6, our empirical work suggests that cultural evolution has manipulated reputation by (1) linking it to kinship and (2) conceptualizing it as transmitted through genealogical lines. Once a population comes to believe that, for example, a son inherits the reputation (in some cases, the sins) of his father, the power of reputation to shape behavior is greatly enhanced.[25]

Cultural Capacities Vastly Increase the Potential for Indirect Reciprocity

Our cultural capacities provide the means to increase both the volume of reputational information available in a population and the accuracy of that information. First, part of our endowment of cultural capacities involves language, and linguistic communication opens up the potential for vast amounts of reputational information to spread around (Smith 2003). This is only part of the story, however, as merely increasing the volume of information won't enable the evolution of cooperation via indirect reciprocity unless that information remains highly accurate. In this regard, it is important to realize two things: (1) individuals will have incentives to inject false information into the system to either hurt competitors (e.g., spread false rumors about someone), or help themselves (e.g., spread good rumors about themselves), and (2) as information flows from one person to another, accuracy will decrease as people misremember, misunderstand, and miscommunicate reputations. Given all this, how can indirect reciprocity persist at all? Unfortunately, most models of indirect reciprocity skip these problems. Cultural transmission provides a partial rescue.

Though not usually thought of in this way, acquiring reputational information from someone about other people is a kind of cultural learning. Consequently, cultural learning mechanisms such as prestige-biased transmission and conformist transmission can improve the acquisition of accurate reputational information for the same reasons that they help in the acquisition of adaptive cultural behaviors. This cultural transmission likely involves observing how other individuals behave toward, and what they say about, potential interactants. This includes both what people intentionally say about someone and how they unconsciously behave toward, and speak about, that person. Our cultural learning abilities are keenly geared to key in on behavior, facial expressions, body language and inadvertent (unintended) comments, which provide a partial check on the intentional verbal statements of individuals who may be attempting to actively manipulate the reputations of others. Combined with this check, mechanisms such as prestige-biased transmission permit learners to acquire reputational information about potential interactants from those individuals who are most likely to have good information. This may involve learning from someone who has spent the most time with the individual in question, or from someone who is judged (by the learner) as best able to accurately evaluate the individual in question. It may also involve giving more weight to the opinions of particularly successful individuals in assessing the reputations of others.

Similarly, conformist transmission allows learners to integrate reputational information received from many people about a particular individual and acquire a more accurate sense of the individual by throwing out the outliers (this helps remove the noise). For example, suppose people judge the reputation of others on a scale from 0 to 100, and five people have mentally assigned scores of 72, 79, 82, 75, and 0 to "Tom." If a learner samples the words and deeds of these five people with regard to Tom's reputation, our learner might estimate the beliefs of the observed judges as 75, 73, 79, 85, and 0. The learner estimated these scores from observations of the five others. Further suppose that the "true" reputational score of the individual is 77 and that the individual who provided the score of 0 completely misunderstood something Tom did. A conformist learner would throw out the 0 and the 85, and assign the individual a reputation score of 76 (by averaging the remaining three estimates). Under a wide range of conditions, this allows a better estimate of an individual's true reputation than averaging across all observations equally. Combinations of prestige and conformist mechanisms are a potent way to improve the accuracy of the reputational information that is the lifeblood of indirect reciprocity.[26] This means that indirect reciprocity will likely be a powerful force in creating cooperation only in highly cultural species (of which humans are the only one known).

Ethnicity and Indirect Reciprocity

When individuals encounter someone they don't know and have never heard of (i.e., someone that they cannot easily get reputational information on), what should they do? Theoretical work on indirect reciprocity shows that they should be SUSPICIOUS: that is, they should defect and see if the other person cooperates. If the other does cooperate, they should switch to cooperating with that person. This is applicable to a variety of circumstances that might lead some social networks to be more densely interconnected than others. However, looking at the empirical work in this book and in the world more generally, these circumstances often occur when members of an ethnic group tend to preferentially interact with each other. Under such conditions, we would expect to see SUSPICIOUS behavior with non-coethnics (because they are outside the reputational network), and NICE behavior with coethnics. Though this observation is important, we do not think it explains the "ethnic bias" in behavior and interaction. Rather, it's a sociological pattern that arises from individuals interacting with others who carry with them both a psychology for indirect reciprocity and a bias for interaction with coethnics (McElreath, Boyd, and Richerson 2003; Panchanathan and Boyd 2003). Below and in chapter 9, we lay out the evolutionary theory that predicts an "ethnic bias."

Costly Signaling and Indirect Reciprocity

Are there any circumstances in which an evolutionarily successful indirect reciprocator should deliver help to (1) a receiver who is not a known cooperator, or (2) a group in which the benefits will be diffused through the group, some of whom may be defectors (this is the n-person case)? The answer is yes (Engelmann and Fischbacher 2002; Panchanathan and Boyd 2003). This can arise if cooperating (and making a big show of it) can have a sufficiently positive effect on one's reputation in future interactions to counteract the immediate costs of helping others, even defectors. This reputation-building form of indirect reciprocity corresponds to another class of evolutionary solutions to cooperation termed *costly signaling*.[27] In this context, individuals are signaling their quality as cooperators to future interactants using reputational effects. Such cooperative acts are especially useful in situations in which many members of a person's social network are present and paying attention. Even if the initial recipient doesn't reciprocate, the giver can have her costs offset by the benefits she receives from the other people who observed her giving and now give to her. Humans should have a psychology that is geared up to look for opportunities of high-broadcast value in which they can obtain big reputational benefits. As we'll see in chapter 8, this effect explains a substantial portion of helping behavior in experimental settings, and it also appears in the real world. Among the Chaldeans, for example, men seize the opportunity to pay for everyone's dinner at a restaurant and argue over who will have the chance to pay. The important thing is that everyone knows who paid. This does not appear to be done with the conscious intent of improving one's reputation; the person truly believes he is being generous solely for the purpose of doing something nice for others—but this does not change the underlying evolutionary basis.

Another implication of this is that individuals should "cheat the system" when they can by seeking out activities and interactions that enhance their reputation at low cost to themselves. For example, among the Chaldeans, being part of a prosocial charity enhances the reputation of its members. Inside the group, however, free-riding in public goods problems reigns because (1) indirect reciprocity is not powerful enough to solve n-person dilemmas within the group, and (2) outsiders to the charity don't have access to information about the relative contributions of individual members. Chaldean social life provides a case study of this phenomenon.

Psychological and Sociological Implications

The available theoretical work on indirect reciprocity allows for a series of general predictions about the nature of human psychology and sociology.

Psychologically, indirect reciprocity predicts the following:

1. People should care about their own reputation and those of others, especially when behaving in the presence of members linked to their social networks. When reputation effects are possible, people should behave more cooperatively than they would otherwise.
2. When interacting with individuals who are not linked to their social network and about whom they have no reputational information, individuals should be SUSPICIOUS.
3. Individuals should prefer cooperative interactions in dyads or small groups (and will likely not cooperate in larger groups), unless an opportunity presents itself for reputation building. This will occur mostly in public contexts because they have the most broadcast value.

Sociologically, indirect reciprocity predicts the following:

1. Dense, bounded social networks that are stable through time lead to the highest levels of indirect-reciprocity-based cooperation. Such networks decrease the chance of meeting someone for whom an individual lacks reputational information and increases the opportunity to cross-check information. In such networks, conformist and prestige-bias transmission can most effectively extract accurate reputational information. These also reduce the proportion of new 'immigrants' who lack proper reputational information.
2. The most effective forms of cooperation via indirect reciprocity will involve dyads. Public goods problems won't be solved by indirect reciprocity except under certain conditions in which broadcast value is high.
3. Culturally transmitted beliefs that extend a person's reputation to her kin will promote cooperation and increase conformity. Such beliefs have the effect of increasing (sometimes drastically) the effects of having a bad reputation by extending them to an individual's close kin, which effectively extends them through time, as when a son gets the reputation of his father. Thus, if a father is considering damaging his reputation in some irresistible defection or norm violation, he faces damaging not only himself but his son and perhaps his son's son. Here "cultural evolution" has taken advantage of our kin psychology to enhance the power of indirect reciprocity.

Social Norms and the Evolution of Cooperation and Punishment

Life in human social groups is regulated by social norms that go beyond cooperation. Norms can be identified by three characteristics (Henrich et al. 2003): (1) they prescribe "proper" behavior (or proscribe improper behavior) for individuals within a population or within some subset of the population (e.g., women must wear veils), (2) these prescriptions are widely shared by at least some significant portion of the population, and (3) failure to adhere sufficiently closely to these prescriptions will anger other individuals in the population (even if the action does not materially affect them), and these angry individuals may take actions that are costly to both themselves and to the norm violator(s). Some examples of norms from anthropology and common experience include eating tabooed food (e.g., pork), going nude at formal weddings, having sex with one's parents, not cutting one's lawn frequently enough, telling jokes during a funeral service, and so on. None of the above theories can explain these panhuman patterns. Why would uninvolved third parties care about what others do, and why would they care enough to take an action that is costly to themselves? And, what does this "norm stuff" have to do with human cooperation?

In thinking about norms, it is important to start with individual minds and then aggregate up (using mathematical models of social interaction and learning) to population-level phenomena. At the individual level, norms begin as sets of mental representations, stored in individual's brains, which we commonly refer to as ideas, preferences, beliefs, values, and practices. These representations prescribe both what their possessor should do in certain situations and what others should do. As prescriptions, norms are often attached to powerful emotions (anger, guilt, and shame) and motivations that lead to strong reactions when the individual himself, or others, violate a norm. These reactions may lead to a variety of forms of punishment, ranging from gossip to banishment and homicide.

At the population level, these mental representations (norms) are shared to some degree, for a variety of reasons, by many members of the same social group. One important reason is that cultural learning mechanisms (like conformist transmission) will lead members of a social group to adopt similar mental representations. A second reason is that the punishment evoked by norms will further lead individuals to adopt practices that avoid punishment, even if they don't truly believe in the rightness of the practice. There are other reasons, but these two classes of mechanisms alone will result in groups' having similar mental representations (norms) about various kinds of behavior. Some of these representations and behaviors have to do with cooperation.

Social norms, despite being one of the most discussed concepts in the social sciences (Bendor and Swistak 2001), have lacked a serious evolutionary explanation that can account for their character and diversity. We think a firm evolutionary explanation for social norms lies in the fact that humans, unlike other animals, rely heavily on their evolved cognitive adaptations for cultural learning to acquire a large portion of their behavioral repertoire, including their social behavior. When both adherence to a norm and a willingness to punish norm violators are influenced by cultural learning, the mechanisms of prestige-biased and conformist transmission can lead to stable situations in which most people acquire and follow the rules, prescriptions, and punishments associated with a social norm. This applies to any norm, be it adaptive, neutral, or maladaptive, and includes norms for costly cooperation (Boyd and Richerson 1992; Henrich and Boyd 2001; Joshi 1987).

This body of formal theoretical work further suggests that although neutral and somewhat maladaptive norms could be maintained within any particular group, group beneficial norms can spread across groups. This happens by competition and selection among social groups that have different culturally evolved norms that vary in their group-beneficial properties, a process termed *cultural group selection*. Furthermore, if these competitions among groups with different norms have been occurring for a long time (tens of thousands of years), the theory shows that the punishment of norm violators within a group will cause natural selection to favor "prosocial genes" (genes that would favor the "high standard of morality" of which Darwin spoke in the quotation opening this chapter).[28] Thus, as a term of reference, we will call the evolved aspects of human psychology derived from this kind of culture-gene coevolutionary process our *social norms psychology* (Henrich 2004a).

Cultural Learning and Punishment Leads to Social Norms

The interaction of cultural learning and punishment leads to norms that are locally stable in social groups. Although the mathematics showing that a combination of prestige-biased transmission and conformist transmission will lead to stable cooperative norms is somewhat more complicated than that used for models of kinship and reciprocity (Henrich and Boyd 2001), the logic is fairly straightforward. Prestige-bias transmission generally acts to favor the imitation of behaviors, values, and practices that lead to the most successful (and often the most fitness-enhancing) behaviors. Operating alone, this mechanism favors *not* cooperating. In contrast, conformist transmission acts to favor the spread of the majority's behavior, and principally acts when the difference in the success between various behaviors or strategies is small or difficult to figure out. That is, conformist transmission can maintain behaviors in a group only if they are neutral, not too costly, or/if their costs are sufficiently ambiguous. Thus, for most important cases of cooperation, or other costly norms, we expect that the costs of doing a cooperative behavior (the "norm") will be too costly for conformist transmission to maintain cooperation, even if it is common.[29]

Adding culturally transmitted tendencies or "tastes" for punishing norm violators can turn the tables on this logic. Imagine a group in which strategies for punishing norm violators and for adhering to a costly norm (e.g., cooperation) are common. Punishing norm violators is costly compared to not punishing norm violators, so many learning and decision-making mechanisms (including prestige-biased transmission) will still favor not punishing. However, punishing norm violators can be substantially less costly than adhering to the norm itself. Here's why: If our group consists mostly of individuals who adhere to the norm and punish those who do not, then most people will stick to the norm (e.g., recycle aluminum cans) to avoid the costs of being punished; if punishers are common, the cost of being punished can easily exceed the costs of sticking to a norm. But if everyone sticks to the norm (because of punishment), punishers don't have to do anything, and being a punisher is not costly. If a few individuals occasionally violate the norm (say, by mistake), the costs to punish can still remain pretty small. Because these punishing costs are substantially smaller than those associated with the cost of sticking to the norm itself (in the absence of punishment), conformist transmission can often maintain the punishment of norm violators against such forces as prestige-bias transmission and experiential learning, even when conformist transmission is not itself strong enough to stabilize the costly norm, without any punishment. How well this trick works depends on the ratio of the costs of punishing to the costs of being punished. The more effective a punishment is (i.e., the smaller the ratio of the costs of punishing to the costs of being punished), the more costly can be the norms that are maintained. In summary: this can work because punishers don't have to pay the costs of punishing very often if being punished is more costly than the costs associated with sticking to the norm—everyone generally sticks to the norm, and punishers need only punish occasional deviants; because the cost is small, conformist transmission can overcome it and keep a strategy of punishing stable in the social group; if punishing norm violators is common, people will tend to adhere to the norm because the costs of being punished for violating the norm exceed the costs of sticking with the norm.[30]

However, suppose that conformist transmission is not strong enough to stabilize the punishment of norm violators. Can costly norms still be maintained? Under such circumstances, conformist transmission may lead to the punishment of individuals who fail to punish norm violators. We call this the "punishment of nonpunishers." By the same logic as above, if a group contains mostly individuals who adhere to a norm, who punish violators of the norm, and who punish those who fail to punish norm violators, conformist transmission can stabilize all of these behaviors by maintaining the least costly behavior—the punishment of those who fail to punish norm violators. This behavior is the "cheapest" way to maintain the norm because those who punish people who fail to punish norm violators will have to pay a cost only when a norm violation occurs (which is rare to begin with) and someone fails to punish that violation (which is rarer still). This means that the costs of punishing need only be paid after the conjunction of two rare events—making it an order of magnitude less costly than punishing norm violators. By favoring the punishment of nonpunishers, the strategy of punishing norm violators is maintained because the costs of being punished for not punishing norm violators exceeds the cost of occasionally administering this punishment, and in turn, the presence of those who punish norm violators means the costs of being punished for violating the norm can exceed the costs of sticking to the norm. In this way, conformist transmission can indirectly stabilize quite costly norms. This may seem odd to some, but as we will discuss in chapter 7, the willingness of people to punish nonpunishers in cooperative dilemmas has now been confirmed experimentally (also see chapter 2).

Actually, the mathematical analysis behind the above idea is even more nuanced. It shows that b (the benefits from the cooperative act) does not influence the creation of a stable norm. That is, the derivation shows that a combination of prestige and conformist biases and punishing behaviors can stabilize any costly norm (c, the cost of the behavior, does matter) independently of whether it benefits anyone. If conformist transmission can favor the punishment of a behavior, then punishment will cause the behavior, practice or belief to stay common in the group. The benefit, b, of the behavior could be zero, or negative, and the group could still "lock in" on the norm. This was an unexpected prediction that fell out of theoretical work on cooperation (Boyd and Richerson 1992), and it may explain the massive database of anthropological findings showing that many idiosyncratic social norms are either neutral or even maladaptive (Edgerton 1992). Although this aspect of our theory explains social norms in general (and cooperative norms are one kind of costly norm), it does not explain why cooperative social norms should be more common, or more likely to spread, than other kinds of norms.

The question of why there are so many *cooperative* social norms can be answered by a process known as *cultural group selection*. As different social groups arrive at and lock in on different social norms, cultural group selection provides a process for selecting among alternative social norms. Some groups will develop norms about constructing community buildings, not eating snakes, and fishing in cooperative units, whereas other groups may culturally evolve norms about cooperatively raiding other groups, sending children to school, and giving young girls clitoridectomies. With these different norms in place, social groups can compete in a variety of ways. First, some groups will have cooperative norms that yield greater success in warfare, and they may spread their norms by conquering other groups (Soltis, Boyd, and Richerson 1995). Second, some groups may have norms that increase their economic or demographic production, so they may spread their norms by generating more carriers of the norms than do other groups. For example, if a group has economic practices that enable them to better feed their children than do neighboring groups, then they will likely have more surviving children to spread their norms. This effect would be magnified if the group also had sexual norms that led to higher reproductive rates than did the sexual norms of other groups (by, for example, making individuals less likely to contract sexually transmitted diseases that produce sterility). Third, if members of different social groups have sufficient interaction (as they often do), prestige-biased transmission can lead individuals to preferentially imitate people from more successful groups, such as groups with a higher quality of life in terms of health, housing, and material possessions. Thus the norms of a successful group can preferentially spread from group to group relatively rapidly (Boyd and Richerson 2002). Each of the above selection processes can lead to the preferential proliferation of cooperative and group-beneficial norms, and each has been observed in the ethnographic, archaeological, and historical record (Atran, Medin, Ross, Lynch, Vapnarsky, Ucan Ek', Coley, et al. 2002; Diamond 1997; Flannery and Marcus 2000; Kelly 1985; Shennan 2003; Soltis, Boyd, and Richerson 1995; Stark 1997; Wilson 2002).

One of the best documented cases of cultural group selection through intergroup competition occurred during the eighteenth century among the anthropologically famous ethnic groups of the Nuer and Dinka (Kelly 1985). Before 1820, the Nuer and Dinka occupied adjacent regions in the southern Sudan. Although the groups inhabited similar environments and possessed identical technologies, they differed in significant ways economically and politically. Economically, both the Dinka and Nuer raised cattle, but the Dinka maintained smaller herds of approximately nine cows per bull, whereas the Nuer maintained larger herds with two cows per bull. The Nuer ate mostly milk, corn, and millet and rarely slaughtered cows, whereas the Dinka frequently ate beef. Politically, the Dinka lived in small groups, the largest of which corresponded to their wet-season encampment. In contrast, the Nuer organized according to a patrilineal kin system that structured tribal membership across much larger geographic areas. Consequently, the size of a Dinka social group was limited by geography, whereas the Nuer system could organize much larger numbers of people over greater expanses of territory. Over about 100 years, starting around 1820, the Nuer dramatically expanded their territory at the expense of the Dinka, who were driven off, killed, or captured and assimilated. As a result, Nuer beliefs and practices spread fairly rapidly across the landscape relative to Dinka beliefs and practices—even though the Nuer were soon living in the once "Dinka environment" and though many Nuer were formerly Dinka who had adopted Nuer customs (so that the differential success can't be attributed to environmental or genetic differences).

Genes Respond in the Wake of Cultural Group Selection

Imagine yourself back in human prehistory, say 55,000 years ago. Social groups are competing, and cooperative norms of various types are spreading through the human species via cultural learning and cultural group selection. Groups have different norms, including those that promote cooperative hunting, the sharing of tool-making techniques, community house building, trade (which requires norms for interaction), raiding, and warfare. The spreading of these different norms effectively changes the selective environment faced by genes because now successful genes have to adapt an individual to a world in which one is punished for norm violations—and the most common norms are often cooperative or prosocial. This culturally evolved selection pressure will favor genes that allow individuals to rapidly acquire the local norms (thereby avoiding punishment), and avoid the temptation of norm violations (i.e., defection). Once the newly selected for genes have spread, cultural learning mechanisms can favor even larger, more costly cooperative norms, and cultural group selection will continue to spread those norms that allow social groups to compete more effectively with other groups (Henrich 2004; Henrich and Boyd 2001). Gradually this interaction of cultural and genetic transmission ratcheted up our sociality and honed our preferences for helping others. Punishing norm violators and avoiding punishment hardened our ability to inhibit quick defections, and refined our learning capacities to promote the efficient acquisition of norms until we evolved into the only ultrasocial primate (Richerson and Boyd 1998). Thus the social norm psychology that underpins much of our contemporary social life emerged.

This theory leads to some general predictions about the nature of human sociality and our social psychology, which we address empirically in chapter 7:

1. Different human groups will be characterized by different social norms, some of which will be cooperative, some not. Some norms will be maladaptive. Noncultural species will not show this kind of variability.

2. Negative reactions by third parties to norm violations will be a human universal (third parties are individuals who suffer little or no cost from the norm violation). Noncultural species will not show third-party punishment of this kind.

3. More costly norms, such as cooperative norms, will involve costly punishment of norm violators by third parties, and possibly the punishment of nonpunishers. This also implies that very low cost cooperative behaviors can be favored by conformist transmission alone, without punishment.

4. Human social groups will vary in their willingness to punish. As we will discuss in chapter 7, when applied to cooperative interactions, this willingness to punish accounts for the empirical phenomena of altruistic punishment and third-party punishment that has been rigorously documented in experimental work.

5. Norms will usually be context specific. Using experimental data, chapter 8 shows how context matters.

6. The strength of norms depends on the ratio of the cost to punishing another individual to the cost of being punished. Thus, social situations that allow for effective punishment to be dealt out at low cost will favor the maintenance of group norms, including cooperative norms.

7. In groups with dense social networks, punishment can operate through damage to reputation. This provides a cheap and effective means to punish norm violators.[31]

8. Cultural beliefs that extend the effects of punishment (such as by hurting the reputation of the punished person's close kin) increase the effectiveness of punishment for norm maintenance. This parallels our earlier discussion of reputation and indirect reciprocity.

Social Norms Can Integrate Aspects of Our Evolved Social Psychology

Building on this, theoretical work is just beginning to show how cultural evolution can take advantage of social norms and aspects of our evolved social psychology in the construction of highly cooperative institutions. For example, by enforcing a norm of truth telling, social norms can enable the transmission of accurate reputational information by punishing those who spread false or inaccurate reputational information. As noted above, this can substantially enhance the effectiveness of reputation-based (indirect reciprocity) forms of cooperation.

Combining social norms and the reputational components of our psychology, recent work has shown how cultural evolution might deploy indirect reciprocity to stabilize cooperation in n-person dilemmas (which indirect reciprocity cannot stabilize, normally). Panachanthan and Boyd (2004) constructed a model that links via cultural beliefs what a person does (e.g., cooperate or defect) in a public goods interaction to her reputational score, which is also influenced by her helping behavior in an indirect reciprocity game (specifically, a "mutual aid game," in which reputation alone can maintain cooperation).[32] Here, once social norms of truth telling have created enough fidelity in reputational information, cultural evolution can harness our reputational psychology to solve the challenging n-person dilemma—which it cannot otherwise solve—by linking it to a dyadic cooperative situation that it can solve. Culture and social norms are essential here in two ways. First, cultural evolution makes a social norm possible that punishes and drives out strategies that (overly) manipulate reputational information, thereby maintaining the integrity of the reputational system. Second, cultural beliefs create the linkage between the mutual aid (indirect reciprocity) game and the public goods interaction by causing both to influence the same reputational score.

Interestingly, as above, this effect can stabilize lots of noncooperative (even maladaptive) behaviors by linking them to reputation in the indirect-reciprocity game. For example, the practice of giving daughters clitoridectomies could be linked to a family's reputation such that families who fail to perform the ritual circumcision would get a bad reputation in their communities, and would thereby not receive aid if disaster were to strike them. This potentially answers the question of how noncooperative and sometimes maladaptive behaviors are maintained in a population by reputation. However, it leaves unanswered the question of why so many societies solve public good problems by using reputation.

As with cooperative social norms above, cultural group selection provides an answer. If different (via the randomness in cultural evolution) groups have linked reputation to a mutual aid game (for example) and to other behaviors—some cooperative (and group beneficial), some neutral, and some maladaptive—then groups that happened to link reputation to cooperative behavior, especially those related to the n-person situations that are otherwise difficult to solve (e.g., cooperation in warfare, defense, and large-scale economic endeavors), will spread relative to other groups via the cultural group selection processes discussed above. Over time we should expect cultural evolution to shape what goes into reputations in a way that benefits the group. Of course, the success of this avenue hinges on maintaining high-fidelity reputational information.[33]

Ethnicity, Norms, and Cooperation

The evolution of norms lays the groundwork for the culture-gene coevolution of an ethnic psychology and the sociological emergence of ethnic groups. Our theorizing begins by considering how the cultural evolution of norms yields new selection pressures on genes. Two problems present themselves. First, individuals have to figure out what the right norm is for getting along in their social groups, keeping in mind that different social groups culturally evolve different norms. By "right," we mean the norm that allows the individual to maintain their reputation, avoid punishment for norm violations, and coordinate their behavior with other members of their social group. Because norms can have an arbitrary character, natural selection will favor genes that direct individuals to learn from those people who are most likely to have the "right norms." The second problem arises because once an individual has adopted some norms, she would be best off to interact with other individuals who share her norms—otherwise she may get punished and/or miscoordinate with those with whom she is interacting. For example, if a man believes that his future wife should have a dowry, he needs to find a woman whose family believes that it should pay a dowry. The root of the puzzle is that people's norms are not stamped on their foreheads. If people knew the underlying norms of all others, they could be sure to interact with those sharing their norms. But underlying norms are more typically hidden properties of individuals that rarely surface, and when they do, it is often too late.

So far, we have focused on norms enforced by punishment and reputation, especially cooperative norms. However, the evolution of our ethnic psychology and ethnic groups can be equally well explained by coordination norms. Unlike situations involving cooperative norms, those involving coordination do not have a free-rider problem. Individuals achieve the highest payoff or fitness by doing what everyone else is doing. For example, the decision to drive on the right or left side of the road is a coordination problem. If you are in England, you want to drive on the left, and if you are in Germany, it's best to drive on the right. Human societies are full of coordination problems, but many of them are substantially more moralized than choosing which side of the road to drive on. Consider the marriage customs of social groups. Some groups demand that a dowry be sent along with their daughter to the groom or to his family. In other groups, the groom or the groom's family (or both) are expected to pay for the bride in, for example, cows, service (labor to the bride's family), or precious metals. Miscoordinations occur when both the bride's and groom's families are expecting payment from the other, and usually everyone loses because no wedding occurs. In a sense, conformist transmission and punishment turn cooperative dilemmas into these coordination situations, so both can contribute to the emergence of an ethnic psychology and ethnic groups.

Now back to the two problems: How does natural selection build a psychology that can (1) figure out who to learn norms from so as to avoid getting the wrong norms, and (2) make sure individuals preferentially interact with others most likely to share their norms? Recent theoretical work on the question (McElreath, Boyd, and Richerson 2003) suggests that selection looks for statistically reliable correlations between observable indicator traits and these underlying norms.[34] Natural selection is fortunate in this case because the same cultural learning processes that transmit the behaviors and beliefs related to social norms also transmit such things as dress, language, accent, dialect, behavioral mannerisms, and food preferences, which are readily observable and tend to be statistically associated with specific sets of norms. This theoretical work predicts that human psychology evolved to seek out "indicator traits" (language, dress, etc.) that match its own because people who have the same markers tend to also have the "right" norms. Using such markers, individuals can bias both their learning from, and interaction with, those individuals who share their same culturally transmitted indicator traits. As above, the proximate psychological mechanism of this ethnic psychology will involve attentional biases (who's interesting?) and affective motives (who does one like being around?). People will naturally prefer to pay attention to, and be around, others who are marked like themselves. These people will often be from the same ethnic group. Thus ethnicity is linked to cooperative social norms because people are using indicator traits (such as dialect) to guide them in acquiring appropriate norms (practices, associated beliefs, and values), and thereby avoid punishment and/or damage to their reputation (which will occur if they adopt behavior that differs from the local norms).

Building on this theory, recent empirical work suggests that people tend to think about certain human groups (those that we'd typically call "ethnic groups") in the same way that humans think about different biological species: we understand the characteristics and attributes of these human groups and biological species in both essentialist (all members carry the same unchangeable essence that gives rise to their shared characteristics) and primordialist (this essence is transmitted down bloodlines) terms (Gil-White 1999, 2001). In the case of ethnic groups, this empirical work suggests that people see ethnic markers or ethnic membership as a cue of hidden, underlying qualities or fundamental (unchangeable) attributes. The three characteristics of a group that tend to spark this type of "species thinking" are: (1) shared markings (shared language, dress, etc.), (2) similarity between parents and offspring, and (3) group endogamy (marriage and mating within the "ethnic" group) and common descent. The greater the degree to which these characteristics are present in a human group—or the greater the degree to which individuals believe or perceive these as present—the more likely individuals are to use species thinking vis-à-vis members of that group. The more individuals use these modes of essentialist and primordialist thinking, the stronger the ethnic bias in interaction and learning, and the greater the importance of ethnic membership on social behavior and cooperation.[35]

Understanding human ethnic psychology and the coevolutionary processes that lie behind it illuminates a large number of puzzling patterns in the world. Perhaps the most general puzzle is why ethnically marked groups seem so important in the world in comparison to other kinds of human groups. Why do ethnic boundaries—and not other possible boundaries—so often mark the fault lines for warfare, genocide, oppression, in-group favoritism, and so on? People could fraction according to height (tall versus short people), occupation (plumbers versus cashiers), or Lions Club membership, but they don't. Why do people often support political candidates that share their ethnic markers? Why is ethnicity so important in marriage, sex, childrearing, and health outcomes? The theory summarized above leads to a series of predictions that help address some of these puzzles:

1. People (children and adults) use ethnic cues to figure out whom to learn from. This kind of learning involves biases in both attention and memory that allow learners to acquire the locally adaptive behavioral norms.
2. People prefer to interact with individuals who share their ethnic markers, to coordinate, and to avoid punishment for norm violations.
3. At a sociological level, these psychological learning biases lead individuals who share ethnic markers to share lots of other norms, beliefs, and values.
4. This psychological preference also creates all kinds of ethnic clumping, as people seek out members of their own ethnic groups in marriage, clubs, religion, politics, and so on.
5. Ethnic markers tend to be hard to fake, as these provide the most reliable cues of the underlying norms. Language and dialect are particularly important, as these cannot be easily learned and can rarely be perfectly faked.

In chapter 9, we apply these ideas to the Chaldeans and show how a little bit of evolutionary theory goes a long way toward explaining a variety of patterns that emerge from the ethnographic data.

Onward

In this chapter we have constructed a theoretical foundation for the rest of this book by presenting the puzzle of cooperation and bringing Dual Inheritance Theory to bear on it. After exposing the Core Dilemma of cooperative inter-action, we have provided a theoretical introduction to five avenues that address this Core Dilemma in various ways: (1) kinship, (2) direct reciprocity, (3) reputation-based indirect reciprocity, (4) social norms, and (5) ethnicity. Moving forward, chapter 4 introduces the Chaldeans by providing some histor-ical and ethnographic background. Chapters 5 through 9 deal with each of these five classes of theory: chapter 5 with kinship, chapter 6 with reciprocity and reputation, chapters 7 and 8 with social norms, and chapter 9 with ethnicity. Each of these chapters proceeds by (1) reviewing briefly the theory discussed above, (2) highlighting some of the key experimental data that bears on the theory, and (3) exploring the Chaldean ethnography through the lens of the theory.

4

The Chaldeans
History and the Community Today

Chaldeans have been in Mesopotamia since 2000 B.C. They are the origins of the ancient civilizations.

—a Chaldean priest talking about early Chaldeans

Terah took Abram his son and Lot the son of Haran, his grandson, and Sar'ai his daughter-in-law, his son Abram's wife, and they went forth together from Ur of the Chaldeans to into the land of Canaan.

—Genesis 11:31

Who Are the Chaldeans?

Many Chaldeans, without much prompting other than a general query about being Chaldean, will spontaneously tell you that they are the source of civilization, the progenitors of the world's three great religions (Judaism, Christianity, and Islam), speakers of the language of Christ (Aramaic), and even the inventors of beer. Chaldeans identify themselves as a group with shared norms, beliefs, history, and nationality. To outsiders, Chaldeans are generally characterized as Catholics from Iraq, although this is an oversimplification. Chaldeans share many Middle Eastern cultural traits yet remain distinct and separate from other Middle Eastern groups because of their non-Islamic faith and their deep historical roots that stretch back into pre-Islamic and pre-Christian times. Although Catholicism is a defining characteristic of the Chaldeans, they are a cultural and ethnic group rather than a purely religious group (although religion is part of the ethnic package).[1] This point was highlighted by the community's movement to have "Chaldean" added as an ethnicity option on the 2000 U.S. census; no religious information can be collected in the census, but ethnic information is allowed. Nonetheless, Catholicism plays a major role in the lives of Chaldeans.

Although many Chaldeans are from lands that are in modern-day Iraq, there are also Chaldeans in other Middle Eastern countries and in India, China, and

75

Mongolia. However, Middle Eastern Chaldeans distinguish themselves from other Chaldeans and acknowledge the non–Middle Easterners as belonging to the Chaldean church but not to the "Chaldean culture." The distinction stems from the geographical location of the person's ancestors, with "true" Chaldeans tracing their origins to the Middle East. The Chaldeans of India, China, and Mongolia were converted in their home countries and thus do not have Middle Eastern ancestry. This conceptualization of ethnicity as endowed by an essence transmitted by descent is a common feature of many ethnic groups (Gil-White 2001), and is consistent with our theoretical framework for ethnicity presented in chapter 3. In Detroit, one consequence of this way of thinking is that the offspring of Chaldeans from Iraq who moved to the United States can still claim full Chaldean status because their ancestors lived in the Chaldean homeland. As we'll discuss in chapter 9, there is no simple definition of who is a Chaldean or what it means to be a Chaldean: it is a combination of descent, religion, language, norms, and tradition.

Most readers likely have never heard of the Chaldeans, although some will be familiar with Tareq Aziz, a Baath Party member and key player in gaining U.S. support for Iraq in its conflict with Iran during the 1980s who became Iraq's deputy prime minister when Saddam Hussein's regime fell in 2003. Interestingly, Aziz's great longevity under Hussein's bloody rule was likely ascribable to his being Chaldean (Christian) and thus essentially ineligible to rule a predominantly Muslim country. As a Chaldean, Aziz could never threaten Hussein's leadership.

History of the Chaldeans

Chaldean History as Told by Chaldeans

Before looking at present-day Chaldeans in the United States, we will briefly discuss the history of this group as told to us by Chaldeans. As scientists interested in explaining contemporary behavior, we are not particularly interested in assembling an accurate account of Chaldean history; rather, we are concerned with the group's culturally transmitted understanding of their own history. It is this history that informs contemporary Chaldean behavior, their views of themselves as a group, and their views of others (such as the many Arabs in Detroit).[2]

Just about any Chaldean in Detroit will explain to you that he is a descendant of the Chaldean civilization of ancient Mesopotamia (the area between the Euphrates and Tigris Rivers situated within modern-day Iraq). More knowledgeable Chaldeans will explain that the earliest mention of Chaldeans dates to the ninth century B.C. when Assyrian King Shalmanassar III indicated that he had encountered Chaldeans, the inhabitants of Chaldea (Bazzi 1998). Chaldeans, as a people, may date back more than eleven thousand years, although it was not until 625 B.C. that the Chaldean dynasty gained political control of Babylon under the reign of King Nabolpolassar. Prior to the seventh century

B.C., Chaldeans had a tribal society with each clan/family under the leadership of a chief (referred to as a "king"). Although the Chaldean era of power was short lived, lasting only 80 years (Kamoo 1999), it was a period of great prosperity and development. By around 538 B.C., the Chaldeans had lost control of Babylon, and after 482 B.C. the area was continually under foreign rule. Nonetheless, during the brief Chaldean era, the civilization restored Babylon and revived the culture of Hammurabi's time, including the restoration of ancient laws, literature, elements of the Old Babylonian government, and commerce as the strength of the economic system. In addition, agriculture improved, animal husbandry increased, and trade and communication routes were enhanced (Kamoo 1999).

Some of the greatest Chaldean achievements came in the area of astronomy. Ancient Chaldeans had a theocratic society and a polytheistic religion. During the Chaldean era, an astral religion developed in which gods were identified with planets. Driven by their religious beliefs, Chaldeans became astronomical experts and used this knowledge, among other things, to develop a seven-day week and a day divided into two-hour increments (Kamoo 1999).

Modern American Chaldeans take great pride that their ancestors (whom they consider to be everyone who lived in ancient Mesopotamia) invented, discovered, or developed the wheel, writing, the first code of law (created by Hammurabi, circa 1800 B.C.), the use of bronze weapons, horse-drawn chariots, irrigation systems and aqueducts, libraries and hospitals, commerce and record-keeping, and the use of an advanced math system involving zero (Sarafa n.d.).[3] They also say that Chaldeans built the Hanging Gardens of Babylon, one of the seven wonders of the ancient world.

Chaldeans see themselves as the progenitors of Judaism, Christianity, and Islam, with the logic going as follows: Abraham lived in Ur of Chaldea, therefore he was Chaldean (Gen. 11:31). Abraham was the father of Jacob, and Jacob was the father of the twelve tribes of Israel, and Judaism. Judaism gave rise to Christianity through Jesus. The virile and vigorous Abraham was also the father of Ishmael, whom he fathered with his wife's handmaiden, Hagar. Hagar and Ishmael moved to Mecca and founded a line that traces to Mohammed, the prophet of Islam. Consequently, Chaldeans believe that all Jews, Muslims, and Christians can trace their roots to the Chaldeans through Abraham. According to a Chaldean priest who is considered an expert on Chaldean history, "Chaldeans are at the foundation of everything important and religious and civil. ... Everything of importance was discovered by the forefathers of Chaldeans" (Jammo 2001).

Origins of Contemporary Chaldeans

An interesting element of Chaldean history, which is not typically recognized by the average Chaldean, is the lack of continuity of the group's name. Though explicit reference to Chaldeans exists in historical and biblical records, there is a gap in time when the group—still residing in Iraq and speaking Aramaic (the language known today as Chaldean)—was not called Chaldean. Disagreement exists as to when and how the group returned to using the name Chaldean, with two competing theories explaining the revival of the name. Both theories agree that the readoption of the name came significantly after the conversion of the group to Christianity. The theories are that (1) the name Chaldean was assigned to the Christians of Iraq by Pope Julius III in 1553 A.D. (Sarafa n.d.; Sengstock 1999), or (2) the religious leaders of the Christians in Iraq chose to be called Chaldeans in order to designate that they are the modern descendants of the ancient residents of Mesopotamia (Sengstock 1999). The gap during which there was no cultural group called Chaldean raises the question of whether there has been cultural continuity (i.e., a Chaldean culture in which the traditions and norms were passed through the generations from the time of the ancient inhabitants to the present). We do know that the Chaldean language has persisted since the Chaldean empire (Rosenthal 1978), and this suggests that it is also possible that other cultural elements passed down through the generations.

The apparent endurance of the Chaldean language is quite remarkable; Chaldean has survived as a minority language for hundreds of years. Prior to Arabic conquests in Mesopotamia, most people in the region spoke Aramaic. However, with the Arab conquest, Arabic was adopted by nearly everyone in the region except for the Chaldeans, who continued to speak Aramaic. Today, all Chaldeans who come from Telkaif [4] (and other neighboring villages) speak Aramaic/Chaldean. They also speak Arabic, which is the official language of Iraq and the language of instruction in school. Chaldeans from Baghdad and Mosul, however, often speak only Arabic. Consequently, in metropolitan Detroit, many of the more recent immigrants speak Arabic whereas the earlier immigrants and their American-born children speak Chaldean. However, many of these recent immigrants decide to learn Chaldean when they arrive in Detroit. Thus, in addition to English, most members of the Chaldean community in metro Detroit speak Chaldean, and the language continues to thrive (although American-born Chaldeans speak it less fluently than their parents). Though it is tricky to trace the Chaldeans and their culture continuously from their ancient ancestors to the present, there is no doubt that the modern Chaldeans constitute a unique cultural group with strong traditions, shared social norms, beliefs, and a well-defined sense of group identity.

Chaldeans Today

Let us now turn to the Chaldean immigrant community and culture as it exists today. All of the data and statements about present-day Chaldeans in metropolitan Detroit come from middle- and upper-class Chaldeans, primarily from the Detroit suburb of Southfield. [5] Chaldeans from Southfield are the target population in our study and constitute the majority of the sample, although some informants live in other socioeconomically comparable suburbs of Detroit. There are, no doubt, some important differences between the Chaldeans of Southfield and those who live in the city of Detroit. Southfield residents have, on average, lived in the United States longer, have more wealth, and show a greater proficiency in English than the Chaldeans of Detroit proper. Consequently, although comparisons of many of the lifeways we observed in Southfield probably can be extended to Chaldeans elsewhere in Detroit, such generalizations should not be made without regard to important economic and demographic variations. Note that the information below on the early Chaldean arrival in Detroit is not limited to Southfield or other suburban communities.

Prior to the last several decades, Chaldeans in Iraq lived in northern farming villages, primarily the village of Telkaif (a.k.a., Telkeppe and Tel Kaif; see Map

Map 4.1 Chaldean Homeland.

4.1), although smaller populations lived in surrounding villages. Each family in the village had a plot or plots of land outside of the village on which it primarily grew wheat and barley and raised sheep, goats, chickens, and the occasional cow. Until recently, Telkaif had a stable population of approximately 12,000 residents, with the population surging to 30,000 during the last quarter century (Sengstock 1999). According to our Chaldean informants, residents of Telkaif tended to marry within the village, so that everyone was somehow related. The villagers supported one another in all domains of life, and households in which three generations lived together and helped one another in running the household and childrearing were common, facts that will become important in the next chapter. Within the last few decades, Chaldeans have migrated from Telkaif to Iraq's large cities, mostly Baghdad and Mosul. In the cities, Chaldeans work in occupations ranging from hotel and store owners to engineers and doctors. Many have become well-educated, attending high school and university.

The Move to Detroit

Migration to Detroit began at the start of the twentieth century. Prior to the late 1950s, Chaldeans came to America in search of economic advancement. Since then, political factors have become a primary reason for immigration, especially after the first Gulf War. Immigration peaked when a loosening of immigration laws in the 1960s and '70s made it easier to enter the United States. Figure 4.1 shows the percentage of our Chaldean informants arriving in Detroit for each

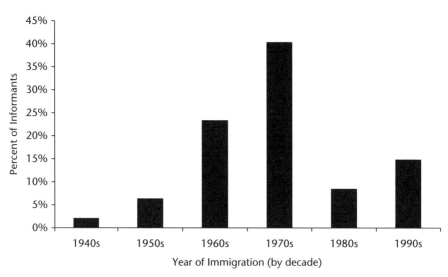

Figure 4.1 Chaldean immigration to Detroit by decade ($n = 47$). Each bar gives the proportion of our sample that arrived in that decade. This plot reflects the randomly sampled Chaldeans used in this study and not the actual proportion of immigrants to Detroit since 1940.

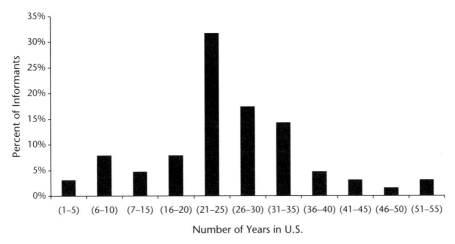

Figure 4.2 Distribution of the number of years Detroit Chaldeans from our study have lived in United States, including both immigrants and American-born Chaldeans ($n = 63$).

decade since 1940. Figure 4.2 shows the distribution of time in the United States for our study community. Most Chaldeans have lived in the United States for about one generation.

Although political motivations are important in moving from Iraq, immigrants continue to be strongly influenced by economic opportunities in America and the desire to reunite with family members who had already come to the States (Boji 2000). In addition, some Chaldeans wanted to leave the country in which they were an oppressed religious minority.

The Economics of Chaldean Immigrants

Early immigrants were drawn to Detroit by the prospect of employment at the Ford car plant. Chaldeans did not end up getting these jobs, however, and instead started working in the grocery-store business. These stores were small mom-and-pop type shops, although many have recently grown into large convenience stores. The stores were, and continue to be, family-run operations that provide employment for their owners' relatives as they arrive from Iraq. Once a few families had settled in Detroit, the Chaldean community rapidly grew as others chose to migrate to a city where they already had relatives. In addition to the economic benefit of being able to work in a family store, there were great social benefits in being around family and other Chaldeans, and a measure of comfort and convenience in being among Chaldean-speaking people. By far, the largest American population of Chaldeans is in metropolitan Detroit, with an estimated 100,000 members. Smaller communities exist in California (primarily San Diego), Arizona, and Illinois.

The neighborhoods in which the Chaldeans clustered varied during the one hundred years that they have been living in metropolitan Detroit. At present, there is a tendency for new immigrants and low-income elderly people to live along 7-Mile Road between Woodward Avenue and John R Street in Detroit; this area was officially named Chaldean Town in 1999. As immigrants establish themselves financially, they often quickly move to the suburbs. Large communities of Chaldeans live in West Bloomfield, Southfield, Oak Park, and Troy. There are five Chaldean parishes in metropolitan Detroit with one in each of the above-mentioned suburbs and one in Chaldean Town.

Family and the Household

Central to Chaldeans' notions of themselves is the importance of family. Relatives in the United States help new immigrants by filling out the immigration paperwork, putting them up in their homes while they search for work and their own residences, assisting with finding jobs (sometimes by having the newcomers work with them, other times by putting them in touch with prospective employers or lending them a car so the newcomers can get to interviews), and orienting them by showing them where the shops are and helping them become familiar with the city. Although most people live in nuclear-family households, non-household family remains important. Adults report seeing close kin, such as siblings and parents, anywhere from once a week to daily. More distant relatives are also seen frequently, especially at church and on special occasions such as weddings. Chaldeans perceive themselves as one large family and are keenly interested in determining how they are related to everyone they meet. Natalie often overheard people being introduced, which immediately led the individuals to start tracing out a kin connection—a process that usually begins by identifying grandparents. Even people who already know each other seem to periodically review how they are related. The feeling of family is so strong that kin terms (such as "cousin") are often used in lieu of names. In one case, Natalie and an assistant were going to an interview with an elderly woman that neither had met.[6] When Natalie asked the assistant for the person's first name, the assistant replied that she didn't know it because when they spoke on the phone the assistant just called her "auntie."

Ties among parents and children, and among siblings, are very strong. Most Chaldeans grow up in lively households with many siblings, although the average size of families is declining. Among Chaldean adults today, the average number of siblings is 4.9—thus, the average number of children in a family is nearly six. In contrast, the average number of children that they have, or plan to have, is 3.2. It will be interesting to see if this number eventually reaches the 2000 U.S. national average of 1.9 (U.S. Bureau of the Census 2004). Despite this drop in reproduction, the importance of, and involvement with, family does not yet appear to be declining. Generally, children do not move out of their parents' house until marriage. Traditionally, people married young, but now that many attend university, they do not marry until their late twenties or early thirties. It is not uncommon for a household to have parents in their fifties or sixties living with three or four grown children, all of whom have careers.

On average, Chaldean households have 4.4 people. Occasionally, a son may be allowed to leave home to attend university, but in most cases sons and daughters must attend local schools and remain in the family home. It is quite rare for Chaldeans to live alone.

The presence of fully grown, gainfully employed children living with their parents raises the average household income, but it is unclear what effect these children have on their household's material quality of life. These patterns are quite variable and, as we'll argue later, in transition. In some families, all members of the household contribute their paychecks to the family, and money and property are held collectively. Some unmarried working women report that when they get married and leave the house, they will turn over all of their property (including money and cars) to their parents—this is simply expected. There are also cases where families immigrated when the parents were older. Often, in these cases, the parents never worked in the United States, and the children entirely support the family. In other families, children do not contribute any money to the household, and parents continue to pay all household expenses—even when the children are professionals in their thirties. Among our sample of Chaldeans, nearly 50 percent of these households had an income in the $50,000–$99,999 range, with the next largest bracket being the $100,000 + range (fig. 4.3).[7]

Although children may not always contribute financially to their parents while living at home, children must care for their elderly parents. Care for the elderly falls primarily to the children, and the community reacts quite strongly if children fail to provide this assistance. Nevertheless, a growing number of Chaldeans are in nursing homes, and in 2000 efforts began to add a "Chaldean wing" to an existing nursing home. Curiously, not one person reported having a parent in a nursing home, although some did admit to having heard that some people have gone into a home. Putting one's parents into a nursing home carries a serious stigma because Chaldeans believe that children should take care of the people who raised them; putting parents into a nursing home shows disrespect. The only acceptable exception is when the parent is very sick and requires extensive medical help that cannot be provided outside of a professional care center.

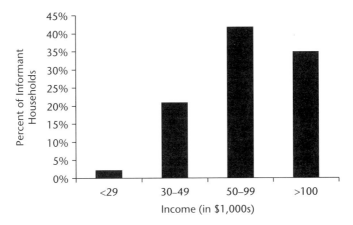

Figure 4.3 Distribution of household income ($n = 44$).

The strength of the family remains intact even as younger Chaldeans grow increasingly acculturated into mainstream American culture. Chaldeans recognize their close family ties, and this is a source of pride and identity for the community. Many Chaldeans distinguish their family networks from those of "Americans" by saying that unlike Americans, Chaldeans care about their families. Sometimes, they would say that Natalie could not understand the closeness of Chaldean families because Americans do not have these kinds of relationships. Regardless of whether their perceptions of other American families are accurate, these kinds of statements reflect the value that Chaldeans place on their family relationships and how they see themselves as different from the majority culture, and thus are relevant aspects of the ethnic psychology that influences Chaldean ethnic identity and group cohesion (which we detail in chap. 9).

Occupations

Traditionally, the majority of Chaldeans in metropolitan Detroit worked in the grocery business. According to the 1962 community census, food and grocery merchandising supported more than half of all Chaldean households. At that time, there were 120 Chaldean-owned grocery stores; this number grew to 278 in 1972 (Sengstock 1999) and reached 1,500 by the mid-1990s (Chaldean Federation of America [CFA] 1999). In fact, almost all of the small grocery stores in the city of Detroit are owned by Chaldeans.[8] This phenomenon is a result of the 1967 riots, after which white store owners fled the city, leaving the area wide open for the Chaldean grocers. Most customers at Chaldean stores are African Americans, and there are tensions between the customers and the owners. Customers apparently feel that these outsiders are overcharging them and that such family-run businesses prevent people from the local neighborhood from being employed at the stores. Crime is a problem at the stores, and several owners and employees are killed each year. When visiting these stores, Natalie grew accustomed to going behind the bulletproof plastic shield that protects workers behind the counter. On several occasions, grocers showed her the guns that they keep on or near them at all times. Periodically, stories would emerge of relatives who had been shot while working in the store, and Natalie was warned several times to leave her car right in front of the door so that she could move quickly from the store to the safety of her vehicle.

Although Natalie heard tales of violence and crime, these were not elements of daily life in the stores. In general, she observed very amiable relationships between grocers and customers. Many of the customers are regulars, and the grocers and customers knew each other's names and would often chat and joke. Many grocers are making an effort to improve neighborhood relations by doing things such as joining the Associated Food Dealers (AFD), which, among other things, sponsors turkey drives at Thanksgiving and cultural education programs so that the customers and grocers can learn more about each other. According to the AFD, stores often hire some neighborhood residents so that the employees are a mix of Chaldeans and African Americans. However, Natalie met very few grocers who employed African Americans. When she encountered stores that employed nonrelatives, the employees were usually Chaldean or, less frequently, Euro-Americans. For the most part, grocers reported good relations with their customers and were quick to attribute any problems to individuals rather than make ethnic or racial generalizations.[9]

Though the grocery business remains an important occupation for Chaldeans, this economic trend is changing. Because many from the younger generations (who were born in the United States or immigrated as children) attend university and pursue professional careers, the number of Chaldean lawyers, doctors, engineers, teachers, and accountants is rapidly rising. And Chaldeans who have come to the United States in the last thirty years often arrived with a university degree, and they enter or resume professional careers rather than joining the grocery industry. A professional in his late forties described the change in the immigration pattern as follows: Until the 1970s, Chaldeans who left Iraq were poor and uneducated, and they came to the United States in search of economic opportunity. From the 1970s onward, Chaldeans immigrated to escape political problems in Iraq or to join family members; these people often were well educated and wealthy.

Our sample of the Chaldean community in Southfield revealed an occupational breakdown as follows: professionals, 45 percent; grocers/wholesalers, 28 percent; unemployed, 14 percent; university students, 11 percent; and non-professionals, 3 percent (fig. 4.4). Our unemployed category consists of women

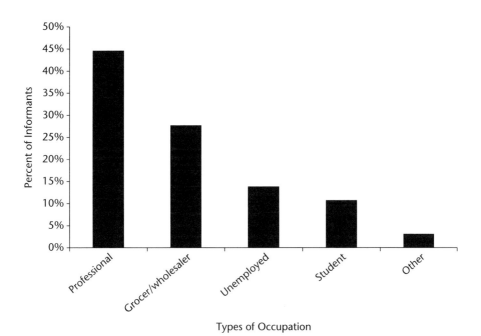

Figure 4.4 Distribution of occupations by type of occupation ($n = 65$).

who are homemakers, retired people, and only one person who is actively seeking employment. Among the professions represented in the professionals category are lawyers, doctors, teachers, financial investors/advisors, engineers, computer-related occupations, psychologists, and real estate agents. Although more Chaldeans are employed in the grocery and wholesale industry than any other single profession, it is clear that there is a strong move away from an ethnic occupation and that Chaldeans are entering a diverse array of fields.

A high percentage of Chaldeans work with family members. This is especially true among those employed in the grocery and wholesale industries, where 94 percent of the people work with at least one relative. A significant portion of Chaldeans in non–retail/wholesale industries also work with family (fig. 4.5). The high rate of related coworkers is particularly noteworthy when we consider that in some occupations the employee has no say in who she works with, such as teachers in the public school system. This means that when a Chaldean has the opportunity to work with family, as is the case with those who own a business or a practice, or is in a position of authority in a company such that he can influence hiring, Chaldeans will frequently seek out a relative as a coworker. The combination of an increasingly educated population, acculturation, and declining interest in working in the grocery business has dramatically changed the occupational profile of the Chaldean community, although the cultural norm/value of working with family remains strong as Chaldeans diversify through the workforce.

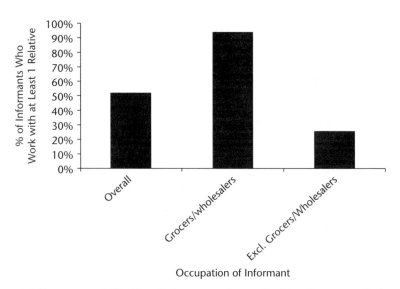

Figure 4.5 Percentage of Chaldean informants who work with at least one relative. *Overall* includes all occupations; *grocers/wholesalers* includes the rate of working with relatives only in these two occupations, *excluding grocers/wholesalers* is all occupations other than these two ($n = 46$).

Organizations

The Chaldean community is very well organized, with numerous agencies and organizations in place to help members of the community. The largest of these is the Chaldean Federation of America (CFA), an umbrella organization that oversees many Chaldean groups. In addition to the CFA, other organizations include the Arab American and Chaldean Council (ACC) and the Chaldean parishes. The community also has social clubs, including Southfield Manor and Shenandoah Country Club, which fall under the auspices of the Chaldean-Iraqi Association of Michigan (CIAM). Chaldean professionals have organized business associations, and leaders in the community actively work to form ties with politicians. Although many of the organizations share a common goal of helping the Chaldean community, competitiveness and tension exists between several of the groups. The ACC and CFA compete for the same pool of government funds, and the CFA is fighting to be recognized as the official organization representing Chaldeans in America. Within the church system, parishes engage in power struggles, and within a church, battles sometimes erupt for control. The Chaldean community faces the challenge of trying to build community cohesion and strength while struggling among themselves over who will bring (and get credit for) this success.

Participation in organizations is extremely high, with 62 percent of Chaldeans involved in a group. To put this in perspective, in 1994 approximately 28 percent of Americans were active in at least one organization.[10] This means that Chaldeans are more than twice as likely to participate in a social group as Americans. In addition to being active at church (either by attending mass regularly, participating in church groups, or doing volunteer work for the church), Chaldeans also focus their involvement on Chaldean-related groups. Of the Chaldeans who have been active in a group in the last year, 82 percent belonged to a Chaldean group—a rate that far exceeds their involvement in any other type of organization. Although participation in Chaldean organizations further strengthens the cohesiveness of the community, it is important to bear in mind that these groups are also a source of tension and divisiveness as they compete for power, money, and prestige.

Summary

Chaldeans generally impress outsiders as warm, hospitable, and hardworking. However, the unity of the community often comes at the cost of excluding outsiders. Most people have no non-Chaldean friends (67 percent) and report interacting with non-Chaldeans only at work or school. Chaldeans prefer to establish business relationships with other Chaldeans as well, and so they tend to use Chaldean lawyers, doctors, bankers, and wholesalers. It may be that the desire to help each other has enabled the success of the community. A new lawyer or doctor need not worry about starting a practice because she can quickly build a large Chaldean clientele, and a Chaldean wholesaler can have hundreds of customers simply by supplying Chaldean grocers. Though some members of the community are becoming increasingly aware of the importance of establishing relations with non-Chaldeans, such as making political contributions in order to increase their influence in government, these goals seem purely practical, and Chaldeans retain a desire to form social relationships within the community. Having said that, we must stress that we were continually struck by the warmth and openness of the community. Almost every person Natalie encountered willingly participated in our study and seemed happy to take the time to meet and talk. The Chaldean community appears to be acculturating rather than assimilating, with the ability to participate and function successfully in mainstream American while maintaining a strong cultural identity and internal cohesion.

Uncited information about the Chaldeans (past and present) is based on statements from Chaldean informants during interviews, conversations, presentations/speeches, or observed interactions.

5

Family First
Kinship Explains Most Cooperative Behavior

We are all related. We are one big family!

—70-year-old Chaldean, speaking of the Chaldean community

Help immediate family, then cousins, and to hell with the rest of the world.

—40-year-old father and lawyer, describing the attitude of Chaldeans

Kinship and families are so much a part of the way we think about the world, it hardly seems a great scientific insight to show that people (in this case, Chaldeans) direct most of their cooperative efforts toward close kin.[1] Nevertheless, a closer look at how kinship actually operates in an ethnic group with a strong in-group bias, such as the Chaldeans, allows us to explore some of the more precise predictions made by evolutionary theory and how our panhuman psychology of kinship influences, and may be influenced by, a long history of culturally transmitted beliefs and values (cultural history). We stress two broad aspects of Chaldean kinship and cooperation. First, Chaldeans in metro Detroit (like all the small-scale societies that have been studied) restrict most of their costly forms of cooperation to close kin, despite an explicit ideology that all Chaldeans are "one big family." In seven domains of costly cooperation, Chaldeans give help to and receive help from individuals who are related to them, on average, more closely than are half-siblings or grandparents. Though this finding is no great shock to those evolutionary scholars working in small-scale societies, it is interesting because (1) the importance of kinship has rarely been demonstrated in a modern urban population, and (2) some evolutionary scholars have argued that the degree of cooperation observed in modern environments may result from a "misfiring" of our kinship psychology. This "Big Mistake" hypothesis argues that people somehow overestimate how related they are to others in their social world because their psychologies evolved in small-scale hunter-gatherer societies. But our data indicate that neither the explicit ideology of one big family nor the apparent unnaturalness of urban Detroit leads Chaldeans to misdirect costly cooperation toward nonkin. Chaldeans clearly know who their close kin are, and they direct costly forms of cooperation toward them. There is no mistake about it. Second, more speculatively, we examine how cultural practices and beliefs related to kinship and cooperation may have combined with our evolved kinship psychology and the economic circumstances of making a living as an immigrant in urban Detroit to create a variety of patterns in the Chaldean community. Perhaps the strongest and most important pattern among these is that one third of all employed Chaldeans work in Chaldean-owned grocery stores (see chap. 4), and Chaldeans own nearly all of the small grocery stores in Detroit. Using the theoretical understanding of kin psychology developed in chapter 3, we consider how Chaldean cultural history may interact with the economic circumstance of new immigrants to Detroit to produce some of the observed patterns. Culturally transmitted norms, beliefs, and practices related to family cooperation, which evolved and adapted to life as a minority ethnic/religious group in Iraq, don't simply vanish upon arrival in Detroit; they influence the decisions and practices adopted by new immigrants and lead to the emergence of group-level patterns of behavior. However, unlike other approaches that view culture as either fixed (and exogenous) or non-existent, ours explicitly models the transmission of practices, beliefs, and values (as mental representations) that may—under predictable circumstances—give rise to varying degrees of stability and change. These transmission processes can be understood as a natural product of our evolved psychology, including our evolved psychology for cultural learning, and explained within a Darwinian framework.

Evolved Kin Psychology

In chapter 3, we discussed how natural selection may have produced a kin psychology in our species (and in many others as well). By way of review: a kin psychology can evolve because (1) blood relatives tend to have copies of the same gene by descent from a common ancestor, and (2) kinship creates numerous opportunities and cues to reliably identify kin. In other species, some of the cues of kinship that allow for kin identification involve scent and proximity. For example, kin may share early experiences that cause them to smell similar, allowing individuals to preferentially direct helping behavior toward others who smell like themselves. And, many animals raise their young together, so individuals can direct preferential behavior at those who were around them when they were young. These types of cues help reduce the problem of determining who will be the recipient of beneficial behaviors. Thus a gene that favors psychological preferences or tendencies (emotions, affective states, etc.) that lead individuals to identify and direct help toward their close relatives will contribute to spreading copies of itself. If you have a particular mutant gene (not on a sex-linked chromosome), there is approximately a 50 percent chance that your brother has an identical copy of the same gene from the same parent.

There is also a 50 percent chance that your father or mother has copies of that gene. More precisely, if a father has a mutant gene that leads him to want to share his food with his children, then there is a 50 percent chance (assuming he's quite certain that he's the father) that he is helping a copy of the gene that resides in his own body. Thus, a father's psychology should allow him to give up some calories to his child that would increase his (the father's) fitness by one unit, as long as the calories increase the child's fitness by at least two fitness units. However, remember, if the father has to give up calories equal to one unit of his own fitness for only a 1.5 unit of increase in his child's fitness, then his preferences, love, and fatherliness should not propel him to do this. Also, recall that relatedness drops off quickly as we move from close family members to distant relatives, so an individual's psychology should be calibrated to reflect this: for example, an individual should be willing to give up one fitness unit for an eight-unit (or better) increase in the fitness of a first cousin. The upshot of this is that kin-based cooperation should be tightly limited to close relatives.

The available evidence largely confirms the above predictions by showing that people use many of the cues used by other species to assess kinship, including physical similarity to themselves and other family members, scent, proximity during youth, as well as culturally transmitted cues like kin terms. These cues influence affective states or feelings that, in turn, influence (along with many other factors) cooperative behavior, trust, and trustworthiness toward cued individuals. For example, in experiments measuring trust using allocations of real money, researchers have shown that people are more trusting of individuals who resemble themselves. The researchers varied "resemblance" by using an image-morphing technology to combine an image of the subject with another person (DeBruine 2002). This suggests that physical resemblance to self may cue affective states built by natural selection to benefit kin. Using the same kind of technology, other researchers have shown that men are more kindly disposed toward babies who resemble them (Platek et al. 2002, 2003). As noted, this is important because men, unlike women, can rarely be completely positive that they are a child's genetic father.[2] A lack of relatedness also explains why adults are much more likely to commit violence (including homicide) against their stepchildren than against their own genetic children (Daly and Wilson 1999). Though not focused on cooperation, a variety of other studies show that our evolved psychology uses scent (Thornhill et al. 2003) and early-life proximity to calibrate affect and target behavior (Wolf 1995).

The Big Mistake Hypothesis

As we said above, some scholars have suggested that much of the cooperation that takes place among nonrelatives in contemporary societies is really an extension of our kin selection and reciprocity-based psychologies (Tooby and Cosmides 1989). In part, they argue that in modern societies our kin selection psychology gets mistakenly applied to nonrelatives and, as a result, people end up cooperating with unrelated members of their group. The Big Mistake argument is premised on two dubious claims: (1) that humans likely evolved in social groups that consisted predominantly of close kin,[3] and (2) that, as a consequence of living in these kin groups, humans tend to assume other group members are close kin. The argument hypothesizes that, in modern societies, where many interactions are ephemeral and involve nonrelatives, our kin psychology may mistakenly cause us to assume, in some sense, that we are dealing with kin and extend some degree of cooperation (Boyd and Richerson 2002). The next chapter deals with another part of the Big Mistake Hypothesis, the reciprocity component.

How likely is it that people mistake nonrelatives for kin? One line of evidence suggesting that people do not generally make this mistake comes from work on incest aversion. It is well established that people have incest aversions (although the reasons for the aversions are debated). However, people do not have a generalized incest aversion to all members of their group, which we would expect if people mistakenly perceived their entire social group as kin. According to the Big Mistake Hypothesis, we cooperate with nonkin members of our social groups because we mistake them for kin. But it seems implausible that we do not mistake group members for kin when it comes to mating but do when it comes to cooperation.[4]

Data from nonhuman primates further challenge the Big Mistake Hypothesis (Boyd and Richerson 2002; Fehr and Henrich 2003). Nonhuman primates are exceedingly competent at identifying kin and selectively directing help at relatives. For example, chimpanzees form kin-based alliances for fighting (Parr and de Waal 1999). However, when primates are placed in artificially large social groups, such as in zoos or research stations, they don't suddenly start cooperating with everyone in the group. Their cooperative patterns remain unchanged: they stick to cooperating with their close kin, despite the novel environment and large group. If other primates can make fine distinctions based on degree of relatedness, including the identification of strangers (chimpanzees kill strangers from other groups; see Manson and Wrangham 1991), it is difficult to believe that humans are worse at this than nonhumans. Below, we show that, despite living in a contemporary urban society and subscribing to a cultural ethos that all Chaldeans are one big family, Chaldeans do not make the Big Mistake.

Chaldeans Restrict Costly Cooperation to Close Kin

Our kin psychology should lead individuals to restrict their cooperation to close relatives. Below, we show that this is in fact the case using data on relatedness for seven different domains of costly cooperation. Across all these domains the average degree of relatedness (r) was 0.33. This means that, on average, Chaldeans direct costly help toward people whose relatedness falls between an aunt/uncle/grandparent ($r = 0.25$) to a full sibling/parent ($r = 0.5$). To provide the reader with some sense of what these r values mean, table 5.1 shows the degree of relatedness between "ego" (some arbitrary focal individual) and his relatives, when there is no inbreeding. In fact, many Chaldeans from the older generations married cousins, so relatedness is, in some cases, somewhat higher than that

Table 5.1 Degree of Genetic Relatedness Between Relatives

Relationship to ego	Degree of relatedness (r)
Parent/child	0.5
Full sibling	0.5
Grandparent/grandchild	0.25
Aunt or uncle/niece or nephew	0.25
First cousin	0.125
Second cousin	0.031
Third cousin	0.0078

given in the table.[5] Remember, the r can be understood as the probability that the other person (whom you may help) carries a copy of the same helping gene that you carry.

Table 5.2 shows that the average relatedness between people interacting cooperatively is remarkably high, varying within the range from 0.23 to 0.43, with an average of 0.33. To construct table 5.2, individuals were asked to list everyone whom they had helped, gotten help from, or would help or ask for help (if the situation arose) in each of the categories listed in column 1. These categories are (1) helping in sickness, (2) allowing someone to live in one's house (or cohabitating in someone's house), (3) providing miscellaneous help (this includes help with daily tasks such as babysitting, shopping, transportation, and practical assistance related to immigration), (4) helping pay medical bills, (5) helping with funeral expenses, (6) providing a business loan, and (7) giving employment. As informants named individuals, we recorded them in the order in which they were stated, and this is captured by the columns in table 5.2 labeled "1st Person" through "6th Person." The final column gives the average r for all individuals listed, regardless of order. In calculating these r values, we used only consanguineous (blood) relationships, and gave affines (non–blood relatives, such as those through marriage) an r of zero, the same as that given to non-relatives. Though this has the effect of reducing the apparent power of kinship (helping your brother-in-law may help your sister, but it is counted as $r = 0$) and

Table 5.2 Average Relatedness of Individuals Involved in Giving or Receiving Help

Domain	1st person	2nd person	3rd person	4th person	5th person	6th person	Overall
Help when sick	0.49 (23)	0.43 (21)	0.23 (20)	0.22 (12)	0.20 (5)	0.13 (2)	0.40 (71)
Getting/giving Housing	0.33 (25)	0.45 (5)	NA	NA	NA	NA	0.35 (30)
Non-$ assistance	0.33 (18)	0.40 (17)	0.29 (10)	0.13 (4)	NA	NA	0.33 (49)
Medical bills	0.41 (16)	0.48 (11)	0.23 (10)	0.23 (8)	0.10 (4)	0.21 (3)	0.32 (52)
Funerals	0.38 (17)	0.38 (12)	0.33 (8)	0.18 (5)	0.13 (3)	0.063 (2)	0.32 (47)
Business loans	0.40 (22)	0.31 (17)	0.25 (13)	0.15 (10)	0.094 (4)	0.094 (4)	0.27 (72)
Employment	0.23 (16)	0.25 (2)	0.25 (1)	NA	NA	NA	0.23 (19)

makes our estimates conservative, the effect is small, as few people actually named affines (which is interesting in itself). Though the importance of the order of naming individuals is difficult to specify precisely, it is worth noting that the average degree of relatedness declines fairly precipitously as we move outward from the first person named, suggesting that perhaps the closest relatives are involved in either more frequent or more salient episodes of cooperation.

To the question about "getting help when sick," 23 informants gave a total of 71 answers: 70 of these responses named people, and one named an institution. Of the 44 individuals who were named either first or second by our informants, only five were not the closest of kin (i.e., 39 were $r = 0.50$). Of those five answers, two were in-laws (i.e., "brothers' wives," which coded as $r = 0$), one was grandchildren ($r = 0.25$), one was a niece ($r = 0.25$), and one was an aunt ($r = 0.25$). This led to an average r value for the first two individuals named of 0.46. However, as with medical bills, the third person named also showed a distinct drop-off in degree of relatedness, with most people naming nieces, aunts, or cousins. After this third person, however, only about half named anyone, and a friend is named for the first time (the first instance of a nonrelative and noninstitution). It's worth noting that there is a clear bias to name female relatives early on. In several cases, close male relatives (brothers, sons, and husbands) are named after aunts, nieces, and sisters. Overall, in the 71 answers, friends were mentioned only three times, and a second cousin was mentioned once; these were the most distant relations.

When asked about whom people would go to for accommodations if they needed a place to live and to whom they would offer accommodations, no respondent gave more than two answers. Respondents were asked separately about giving versus getting accommodations, and in the analysis, the first responses to each question were averaged, as were the second responses. Twenty-five people gave a total of 30 responses to the two questions, with an average degree of relatedness of $r = 0.35$ for the people named. The average degree of relatedness was higher for the second person named ($r = 0.45$) than for the first person named ($r = 0.33$). This occurred because some of the first people named were relatives of the respondents' spouse, and these people (usually) have a degree of relatedness of zero. Since no one mentioned a spouse's relative as a second response, the average relatedness for second responses was higher than for first responses. It is important to remember that even though the spouse's relatives have no genetic relatedness to the respondent, they are genetically related to the spouse, and housing is being provided/offered to both the respondent and the spouse.

In the category of nonmonetary help (babysitting, shopping, transportation, etc.), all but one of the 18 informants readily gave two answers, and one gave five—yielding a total of 49 answers. The average degree of relatedness for the first two responses was 0.36, with 21 of the 35 answers being the closest of kin. The remainder included in-laws, uncles, grandmothers, grandsons, and cousins, but no friends—in fact, friends are never mentioned in all 49 answers. The number of people answering dropped by half for the third person named, and second cousins appear for the first time. Among the four informants who named a fourth person, one named a third cousin.

Sixteen informants gave a total of 52 answers to the question about getting help with medical expenses. Of these 52 answers, nine were institutions (the church, a Chaldean charity, the government, etc.) and three involved asking the doctor to provide either free medical care or a payment plan. Of the people listed first by our informants (i.e., all responses other than institutions), the three who wanted help from the doctor were the only ones whose coefficient of relatedness to the person being asked for help was not 0.50—everyone else listed either a parent or sibling first. Of those individuals listed second, all were parents or siblings except for one person, an uncle. The average degree of relatedness was higher for second responses than first responses because of the people who mentioned doctors—doctors have (usually) no kin relatedness to the respondent, and all mentions of doctors were given as a first response. By the third and fourth person listed, the kin circle had drastically expanded, but about half of our informants stopped naming anyone. Of the responses from those who did name someone, the average r value was 0.23. This circle included mostly cousins, although a friend was mentioned by one person. By the fifth person named, spontaneous responses came from only four informants, and half of those did not even consider people with an r of less than 0.25. These circles included friends, third cousins, nieces, and uncles. The average r value of the sixth person listed is higher than that of the fifth person listed because one of the three informants named her sisters. Undoubtedly, close kin are the first people, and in some cases the only people, who come to mind when thinking about getting medical bills paid. In considering the relatives named, the reader should keep in mind that, unlike many other Americans, most Chaldeans can name at least up to their third cousins, and some can name their fifth cousins.

In the case of funerals, 17 informants gave 47 answers. Of these 47 answers, 17 were "the box," referring to the donation box that Chaldeans put out at funerals, and only one was an in-law. The box allows people at a funeral to donate money in an envelope, usually marked with their name, to the family of the deceased. Of the 17 people named first, 15 were the closest kin, and one was the spouse of a daughter. What may matter here is not only the relatedness to the living who need to pay for the funeral but also the relatedness to the deceased. For example, one woman explained that she went to her aunt for money when her mom died. The informant is related to her aunt with $r = 0.25$, but her mom was related with $r = 0.5$ to the aunt. By the third person named, the number of informants giving answers dropped to half, and both the average value of r and the number of informants responding dropped precipitously after this. In all 47 answers, friends were mentioned only twice, and the most distant relatives mentioned were first cousins.

In the category of "business loans," 22 people were interviewed, and they listed between one and six people they would go to (or have gone to) for a loan, yielding a total of 82 answers.[6] Of these 82 answers, only 10 were "the bank" (88 percent of answers were not formal lending institutions). Eight of these 10 "bank" answers were given first. These 10 responses were removed in constructing table 5.2 so that the table reflects only individuals to whom a person would go for a loan. Focusing only on people's first answers (the first thing to come to mind), not only is the average $r = 0.40$, but 16 of the 22 nonbank answers were first-order relatives ($r = 0.50$, siblings and parents). The only nonconsanguineous kin mentioned were in-laws (affinal kin, $r = 0$), who were mentioned four times; "friends," who were mentioned once; and a prestigious Chaldean who was mentioned by name once in the "named first" list (and four times overall). In the "named second" category, 18 people (of the 22 interviewed) freely responded. One person listed "the bank," and the 17 others yielded answers with a mean of $r = 0.31$. Of these 17 answers, two were in-laws, three were the same prestigious Chaldean, and the rest were a variety of blood relatives. Friends and cousins of various types do not begin to enter people's answers with any frequency until the third and fourth names listed.

Our probes regarding employment, though still eliciting fairly high degrees of relatedness, showed substantially lower overall relatedness than our other questions. There's a simple reason for this: people tended to think about who might actually be in a position to give them employment, so this question yielded answers that combine kinship and the possession of a particular kind of resource. The 16 informants for this question produced only 20 answers, one of which was "Ford Motor Company" (and which we did not include in table 5.2). Of the 19 answers involving people, two were friends, three were in-laws, and 14 were blood relatives. The most distant such relative was a first cousin, and most people named only one person.

These inquiries show a consistent picture: despite their wide social networks of other Chaldeans, their knowledge of distant relatives, and their ideology of one big family, Chaldeans confine costly cooperation to kin, and the closest of kin are consistently the most salient. For the most part, they don't even think of naming people who are not close kin, and the precipitous drops in the number of persons named and the degree of relatedness qualitatively reflects the threshold nature of Hamilton's Rule.

Ideology, Practice, and the Big Mistake Hypothesis

One of the most interesting things about the high relatedness between individuals engaged in cooperative relationships is the accuracy with which Chaldeans direct their assistance to close kin—a finding that is clearly at odds with Chaldean cultural ideology and perceptions. When Chaldeans speak about their community, they often emphasize that almost everyone's ancestors come from the village of Telkaif, where all the residents intermarried, and that everyone is related, even if very distantly. One priest said that when a new immigrant arrives, he looks for work with a relative, even a distant relative whom he has never met, because their grandparents will have known each other in Telkaif. In reality, our data show that people do not tend to look for work with distant relatives; rather, the average degree of relatedness between people who give a relative employment and the employee is 0.354—a closer relative than an uncle or grandparent. Hardly a distant relative! Our work reveals no case of someone coming and working for a distant relative, so if this happens, it's fairly rare.

Given that Chaldeans seem strongly wedded to the idea that they are all related, one might expect that prosocial, cooperative behavior toward other Chaldeans could result from a kind of misfiring of their kin psychology, leading them to confuse close kin with distant kin and nonrelatives. The above findings, however, do not support this view. People's actual costly cooperative behavior reflects the reality that they are not one big family, and demarcates (with remarkable precision) the lines between close kin and the rest of the Chaldean community. As Table 5.2 shows, there is a consistent drop off in the number of individuals naming distant relatives once relatedness goes beyond uncles and first cousins. Not only do Chaldeans selectively direct help to close kin, they also prefer to receive help from close relatives. A man with ten siblings said that if he was sick, he would want help from his immediate family ("the immediate family is capable of doing it"). And when it comes to borrowing money, Rose explained that a close Chaldean friend offered her money, which she needed, but she turned her down—she preferred to take money from her brothers (although she did accept food from the friend). Since the kinship psychology is applied so accurately to close relatives in the urban ethnic context of Chaldean life, it seems unlikely that group identity and seemingly group-related cooperative behavior results from kin psychology writ large.

Nevertheless, we don't think that all the one-big-family talk is just meaningless cultural fluff or cheap talk. The Chaldean myth of relatedness clearly promotes a sense of ethnic identity and cohesion, which likely leads to a preference or bias towards helping, interacting with, and learning from other members of the Chaldean community (such as working and socializing with other Chaldeans)—but not because of any connection to a kinship psychology. Instead, in chapter 9, we will argue that this ideology arises and interacts with our ethnic psychology.

Kinship Organizes Business

Chaldean families help each other in nearly every aspect of their lives. The most obvious cooperation occurs in the family-run convenience stores, where parents and children work together. In addition to nuclear-family stores, two or sometimes more brothers often jointly own a store or stores, and their wives and all the children work in them. In our study, 33 percent of employed Chaldeans work in the retail or wholesale grocery-store business as their primary occupation, and 94 percent of these people work with at least one relative. Children learn and develop in a context in which working in the family store is part of the normal routine of daily life. Although there is an emerging trend for Chaldeans to choose other careers, most continue to help part time in their parents' stores. This expectation places costly time demands on the children (and provides few direct benefits, such as good wages). For example, a man in his mid-twenties, who works full time in finance and attends university in the evenings, still finds time to work at his family's store on the weekends and whenever else he can manage. He did not view the extra work at the store as a burden—it was merely the norm, as such schedules are common among many young Chaldeans. Similarly, a female accountant in her twenties works part time at the family store in addition to attending college; she used to work at the store full time. She explains: "Working in the store was normal. I never thought about it or questioned it." Another professional, now in his early thirties, described how his grandfather opened a grocery store when he came to the United States in 1929; his father also went into the store business, and at age 5 he started working in the store on Saturdays. In high school, he worked three or four days a week in the store, and when he went away to university he would return home in the summers to work there.

   The cultural learning theory discussed in chapters 2 and 3 provides some theoretical tools for understanding these ethnographic patterns. Two aspects of our learning psychology seem important here. First, humans seem to rely on conformist transmission in situations where there is uncertainty about what the best thing to do is, or when necessary information (to make an evaluative decision) is difficult or costly to obtain. Second, we also use ethnic cues to figure out whom to pay attention to for the purposes of social learning. This means that in deciding how much to work in the store, young Chaldeans may be sampling their fellow Chaldeans (matching by age, gender, and success), and not Americans in general, to see how much they should work in their parents' store. By using their innate learning abilities to address the problem of how much to work in the store, young Chaldeans would conclude that spending substantial amounts of time in their family's store is appropriate. Numerous cases show that economic incentives are not explanatory, as the professionals who continue to work in the family store on weekends could earn more by putting additional time into their respective professions and paying wages for someone to work in their stead. But this is simply not done. These same cultural learning mechanisms can provide the dynamics for changing occupations, as individuals on the outskirts of Chaldean social networks (where conformist transmission has little influence) or without store opportunities focus on and learn from prestigious non-Chaldeans.

Why Family-Run Grocery Stores?

An integration of economic, cultural, and evolved psychological factors helps explain the persistence and character of the Chaldean's pattern of grocery-store ownership. Running the stores as a family business produces several economic advantages. With the inflow of new Chaldean immigrants, grocery stores provide a crucial source of employment for newly arriving family members. This situation provides trustworthy cheap labor for store owners and an essential social and economic entrée for arriving immigrants. Many immigrants tell the same story of arriving in Detroit and immediately starting to work in a relative's store. Even when the new immigrants are highly skilled professionals, they often work in the family stores because their credentials are not recognized in the

United States. Until very recently, families also employed relatives who were not legally allowed to work in the United States. This was often the case when Chaldeans came to this country on student visas, which do not permit employment. In many of these cases, the students enrolled in college, worked illegally in a family store, and eventually married a Chaldean with American citizenship. The mere fact that many new immigrants don't speak English provides yet another advantage to working for relatives who speak their tongue. Store owners also get advantages, as revealed when they explain that employing family members enables them to "trust" their employees. The stores are a cash business, and some feel that the ease with which employees can steal necessitates working with family because you can "trust that kin won't steal." Though many successful businesses run without employing family members or relying on kin-based trust, on a small scale,[7] there are no doubt large efficiencies to be gained from trustworthy employees, and our kin psychology provides a strong foundation for this kind of trust. Moreover, it seems that Chaldeans, perhaps because of their cultural heritage, seem less trusting of nonkin and strangers than do typical Euro-Americans and thus are even more inclined to rely on family.

A final reason for the pervasiveness of family-run grocery stores among the Chaldeans, which was not explicitly stated by store owners, helps explain the attraction to the business for new immigrants and the Chaldeans' overall dominance of this business niche in metro Detroit. Chaldean notions of family and their culturally acquired emphasis on economic contributions to a joint enterprise allow these businesses to outcompete many potential non-Chaldean competitors by paying low wages to relatives (often below minimum wage), pool profits for further investment (rather than paying them out in health insurance and wages), avoid costly security and accounting measures (needed when nonkin are employed), and use family as a flexible labor pool to match demand. Chaldeans' cultural preferences or beliefs—their tastes for family-based forms of economic organization and their distrust of nonkin and non-Chaldeans—combine with the competitive viability of family-run stores to generate many economic successes. In the wake of these economic successes, new immigrants, using their cultural learning mechanisms (ethnic, prestige, and conformist biases) to figure out what to do in Detroit, frequently acquire the notion that family-based store ownership is the path to success. This led to the proliferation of Chaldean-owned grocery stores. Furthermore, Chaldean families who, for whatever reason, were unable to maintain the family-based administration of the stores may have had less economic success (at least at first) and would not have been used as role models by new immigrants.[8]

Ethnographic interviews fit this theoretical model. Typically, after a new arrival has worked for a relative and learned the business, adjusted to life in the United States, and earned a little money, he leaves the relative's store to open his own shop. The most commonly cited reason for owning a store is that the person learned this occupation when he worked for his relatives upon arriving in the United States. As one grocer put it, "Everyone had stores, so I just joined the party." Arriving Chaldeans seem inclined to take up the occupations of other successful Chaldeans.

Both kin psychology and cultural learning are important in understanding the above case. Kin psychology provides the foundation for the kind and degree of cooperation that is necessary for running Chaldean stores. Cultural learning adds three important elements. First, the Chaldeans' particular beliefs and values about family cooperation in joint economic enterprises means that they arrive culturally prepared for entering family-run businesses; in Telkaif, farms are essentially family-run businesses. Second, the economic and linguistic circumstances of arriving in Detroit mean that new immigrants learn the skills and knowledge associated with running grocery stores in Detroit ("human capital"), and, as a result, these individuals are more likely to be successful at operating grocery stores than they would be in, for example, plumbing, masonry, or aerospace engineering. Third, aside from the skills learned upon arrival, immigrants use cultural learning capacities to figure out what to do once they are settled. Initially, as noted, the success of the family-store niche in Detroit would have led new immigrants to copy successful coethnics (using ethnic and prestige biases). Later, this effect, which may have decreased as the competition among the many Chaldean stores increased, would have been augmented by a conformist effect, given that most of the Chaldeans that a new immigrant meets are members of store-owning families.

In most American cities, some economic niches, such as small grocery stores, are filled by one or another immigrant group, who use some form of kin-based economic organization often to great or moderate success. For example, in various cities, many Korean families own corner grocery stores, Chinese families own dry cleaners, Vietnamese families own nail salons, and Indian families own hotels and motels (Booth 1998; Hua 2002; Park 1997; Sanders and Nee 1996). We suspect that a theoretical architecture of the sort we have presented, one that combines kin psychology with cultural learning, could be used to explain this robust phenomenon.

Beyond the Grocery Store

Family cooperation in economic domains extends beyond the grocery store. In many occupations, employees do not influence hiring decisions. Accordingly, we do not expect to see family working together at large companies, such as Ford, or in government institutions, such as public schools. In these situations, Chaldean relatives should be no more likely to work together than people from any other cultural group. However, when a situation occurs in which a Chaldean does have the opportunity to hire, or work with, a relative, he is quite likely to do so. Even in nongrocer professions, 26 percent of Chaldeans work with at least one relative. Though we don't have data from a comparable group in Detroit, we strongly suspect that this is extremely high compared to Euro-Americans of comparable class and geography. In some of these cases, siblings or husbands and wives have gone into business together, and in other cases a Chaldean has a senior position in a company and is able to influence hiring (and usually hires a cousin). Though preferring to work with family is not necessarily cooperation, it suggests that the preference to work with kin may triumph incentives to pick the most talented, smartest, and best-trained working partners, as usually those won't be your family members.

As we saw above, economic cooperation within Chaldean families extends far beyond working together. Relatives cooperate economically by lending money, usually without interest, and by co-signing on bank loans. In contrast to the more typical American middle-class practice of getting business loans from banks, only 24 percent of Chaldeans say that they would go to a bank as either their first or second choice. Also interesting is the fact that several Chaldeans said that they would ask a particularly wealthy man in the Chaldean community for assistance. This man, a wholesale distributor, helped many of the Chaldean grocers get started by lending them money, with interest, and requiring that the grocer buy products from him. He no longer makes loans (we asked him), but many people are unaware of this and still claim that he would be a good option for aid.

Cooperation in Noneconomic Domains

Chaldeans provide an extremely important form of help when family members first arrive in the United States—and even before. Siblings, parents, and in-laws provide aid prior to the immigrants' arrival by sponsoring their immigration and helping fill out paperwork. Once the relatives arrive, they usually live with a member of their immediate family for anywhere from a few days to a few years. Michael, who was accustomed to living alone, at one time had 11 relatives living in his home. His sister came to Detroit and moved in with him; three years later, his other sister and her family arrived and moved in with him for two years, and his brother and his wife and children arrived and moved in for three years. Although family is willing to help each other, they are aware of the costs. "Helping doesn't come free," Michael said. "The cost is lost privacy." Kinship does not mean people are blind to self-interest and the costs of cooperation.

Family also helps with employment, often by bringing the new arrival to work in the family store, by lending a car so the new immigrant can search for a job, or by driving the person from job interview to interview. Relatives and friends from Iraq who are already settled in the area help with acclimatizing new immigrants by, for example, showing them where stores are located, enrolling the children in school, and helping them learn their way around the city. A grocer who came to the United States at age 21 expressed a frequent sentiment when he said that his family helped with "everything from A to Z." In his case, his parents-in-law did his immigration paperwork, put him up in their house for six months, gave him work in their store for three years, and then set him up with his own store. Later, he brought his family over from Iraq, and his brother lived and worked with him.

As shown in table 5.2, relatives also rely heavily on each other in times of illness. When a person is sick and needs help around the house, transportation to the doctor, and companionship, Chaldeans turn to their close kin. Although people help an extensive family network, including aunts, cousins, and grandparents, most assistance takes place within the immediate family. Chaldeans are more likely to turn to their closest kin when they are sick than for any other need. The type of help that a person offers depends on his or her gender. Women help with the cooking, cleaning, and groceries, whereas men tend to help with transportation. Sue told me that when her mother was dying, she and her sisters cared for their mother—apparently her brothers were of little use because "they can't even make tea!" People are clearly hesitant to ask friends and distant relatives for help because they don't want to bother or impose on them and because they believe this is the responsibility of immediate family.

Other types of cooperation within the family include lawyers giving free legal advice to close relatives and discounted advice to more distant relatives. One lawyer gives free advice to her siblings and aunts/uncles, with discounted advice given to cousins, whereas a second lawyer gives free advice to siblings and first cousins, with discounted advice to second cousins and his siblings' friends. Once again, we see that people are keenly aware of their relatedness and adjust for it. On a day-to-day basis, women help their siblings, children, and cousins with babysitting; some women drive their sisters or mothers to appointments, funerals, and the grocery store, and men help relatives with moving or household repairs and renovations.

Changing Values

The Chaldean conceptualizations of communally owned family property and labor bear a strong resemblance to patterns of kin relations observed across the human spectrum (Brown 1991; Fiske 1992), despite being somewhat different from those of the middle-class European Americans that now live interspersed among the Chaldeans. Chaldean beliefs and practices in this regard appear to be shifting from an emphasis on communal property and the good of the household to an emphasis on individual ownership and high levels of investment in children. For more traditional Chaldeans, economic help flows in one direction—from parents to children—when parents have young children, and the pattern reverses itself as the children grow older, with children supporting parents, or children and parents contributing collectively to the household. Lida, referring to the paycheck that she turns over to her mother, said, "It is not my money, it is the family's money." There are also families that communally own property. In some families, cars do not belong to any one individual, and when a child moves out, the car stays behind. In other families, everyone owns the house and family store collectively, although only one person may have his name on the mortgage and other documents. In these communal cases, the family acts as a single economic entity that functions cooperatively by giving up individual economic gains to increase the combined prosperity of the family.

In contrast to these highly communal families, parents in other Chaldean households fully support their children until they move out. The children, who frequently work full time in professional careers, save their money for when they marry and move out. Cars, some of which are bought for them by their parents, are theirs to keep when they move. When Chaldean males get married, they expect to buy their own house.

It appears that the number of economically cooperative ("traditional") households has been declining over time and that the current communal families waver at an unstable equilibrium from which the other households have slipped away. We suspect that the communal model of families results from a combination of our evolved kin psychology and culturally acquired notions that many Chaldeans brought with them from their farms in rural Iraq. We further speculate that these cultural beliefs and the communal model served early immigrant families well. When Chaldeans were recent arrivals in the United States, parents may have needed the extra income that the children brought in so that they could afford to move from Detroit to the suburbs, buy cars, expand their stores, and so on. A 27-year-old man with ten siblings explained that his parents were already old when they brought his family to Detroit, so the grown children supported the family. He described the money and property of the immediate family as "all one pocket, one fund." When families move to the suburbs, both children and adults observe successful and prestigious individuals, and as we discussed in the introduction, people tend to acquire their ideas, values, beliefs and practices of prestigious individuals. In a middle-class environment, where resources invested in skills and individual wealth (marked by cars, watches, houses, clothes, etc.) allow individuals to acquire the markers of success and prestige, families who invest heavily in their children will (inadvertently) perpetuate their practices. These processes will gradually lead to the decline of the communal model and the spread of the model of heavily investing in offspring that characterizes much of the American middle and upper class (Richerson and Boyd 2005).

One could argue that this change in family strategy (investment) results from some kind of rational cost-benefit analysis related to a family's change in wealth as they move from the city to the suburbs. The argument suggests that older children supported their parents and siblings because there was no other option. It may be that cooperative finances is a survival strategy that families use when they are poor, or at a time when they are trying to make the transition from lower to middle class and need the extra money to boost them over the boundary. As monetary pressures decline, parents may be more likely to take on the full financial responsibilities of the household, as is common in other middle-class American families. However, three interrelated factors make such an explanation difficult to maintain. First, many middle-class Chaldean families are still maintaining the communal style; they have not switched, and they seem firmly committed to the communal model. If cost-benefit analysis were the sole mechanism operating, we should expect all or most families to switch rapidly and without any emotional commitment. Notions about the "right way" to run a household would be like deciding to put one's boots on when it rains, and sandals when it's sunny. Second, the shift is occurring on the kind of generational time scale that characterizes cultural learning processes, not rational cost-benefit analysis. The switch is occurring, but it is taking generations, not minutes, days, or weeks. Third, it's hard to imagine how individuals would obtain the information needed to make the required cost-benefit calculations, given that they cannot (and do not) do experiments and have no prior experience with the practices they are adopting. More generally, such cost-benefit models typically lack psychological plausibility and fail these kinds of reality checks (Henrich 2002). Nevertheless, the nature of our data in this specific Chaldean case prevents any decisive tests, so we offer the above discussion of the changing values of children as speculative, although informed by other direct tests of cost-benefit approaches.

The Influence of Chaldean Cultural History on Cooperation Today

Although the theoretical foundations of our evolved kin psychology help us greatly in understanding the patterns of costly cooperation among Chaldeans in metro Detroit, a more complete understanding is provided by considering that humans, unlike other animals, are heavily reliant on cultural learning for acquiring large portions of their psychology and behavior (chap. 2). Although all humans have a kin psychology, this theoretical concept alone cannot account for the high levels of variation in behavior toward kin that emerge as one looks across the human spectrum or for the changes in patterns of behavior over time (chap. 3). Parents, their offspring, and full siblings all usually share the same degree of relatedness, and yet the amount of cooperation observed in these relationships varies across behavioral domains and across social groups. Further, as we observed above, the details of when and how much help kin should give each other change over time; a good example of this is the trend for Chaldean children to contribute less to household maintenance as more mainstream American middle-class beliefs and practices flow into Chaldean households.

Standard views of kin psychology can be augmented with an understanding of cultural learning. By thinking of kin psychology as (1) predisposing individuals to help and be helpful to their close kin, and (2) favoring the acquisition (cultural learning) of ideas, beliefs, values, and practices that benefit kin, we can create a more complete understanding of kin-related behavior that is capable of explaining variation among groups and changes through time. Though this will favor the acquisition of ideas and practices that can benefit kin, the transmission of this cultural stuff is also influenced by other kinds of transmission biases (such as prestige-biased and conformist transmission) and by patterns of social interaction and social structure that influence the contact between different individuals. Economic factors with direct effects will influence decision making directly, whereas more subtle economic factors will influence the spread of cultural beliefs (e.g., certain ideas may lead to more success, which will make the idea more likely to be imitated) over longer historical times scales. As we discuss below, culturally learned norms, ideas, and practices related to economics (farming) or social structures (who lives where) can influence the amount and kind of cooperation among kin. Cultural norms and beliefs, unlike physical environments or buildings, are carried in people's brains and taken with them when they move to a new place (e.g., Sowell 1998). Although many of these learned behavioral patterns may be inapplicable in the new locale (planting wheat is difficult in Detroit), others (e.g., family values or organizational forms) may change only after generations of cultural learning and adaptation.

Though all human groups focus their most costly forms of help on kin, the amount of help and the way that their help permeates all realms of their lives set the Chaldeans apart from many other urban Americans. Whether such high levels of cooperation are common among other American ethnic groups would be interesting to investigate. Although it may be the case that minority populations, primarily those who are relatively recent immigrants to the United States, have more within-group cooperation than other American communities, we suspect that Chaldeans would still fall at the high end of the cooperation scale. (we base this, in part, on Natalie's research among the Hmong of Detroit). This begs two questions: Why do Chaldean families cooperate more with relatives than do other cultural groups? And, why does cooperation within Chaldean families take different forms than cooperation among non-ethnic American families? Some of the answers may lie in the recent cultural history of the Chaldeans, during which time the physical, social, and economic environment of life in Iraq influenced the cultural evolution of certain kin-oriented cooperative norms, beliefs, and practice, which immigrants carried with them to Detroit.

For much of the period since Chaldeans adopted Christianity, they have lived in small farming villages in northern Iraq where people reside in extended family households. The ideal household arrangement was a man and his wife (who were preferably cousins), their unmarried children, and their married sons with their wives and children. As we said, most families lived in a village and maintained small plots of land on its outskirts. All the males in the family worked their land together while the females took care of the household and the children. Within this context, members of a family were mutually dependent on each other, and their livelihood required the efforts of all. The need for multiple males to work the fields supported the household structure of sons' remaining in the family home, or at least living nearby so that the family could function as an extended household. The circumstance of economic life would have favored those families that possessed norms, beliefs, and values that assured maximum cooperation among family members and the maintenance of the family unit (children would have acquired beliefs and values about staying home). The success of these arrangements would have led to greater prestige for the families that possessed these practices and the proliferation of these beliefs, values, and practices to other families. However, kin psychology would have prevented this kind of cooperation from spreading outward beyond close relatives to other individuals without additional mechanisms; in this case, kin psychology acted to break the dissemination of values that would extend high levels of cooperation and help to nonkin or distant kin.

Moreover, once these practices are common, a number of mechanisms provide some cultural inertia, which act to slow the intrusion of competing ideas even in economically altered circumstances. First, conformist transmission will act to impede the entrance of novel ideas. Second, cultural learning processes tend to interweave cultural ideas together such that although novel ideas may increase economic success, they may not fit with or complement other existing beliefs. Third, once behaviors, ideas, and practices become common, they are often moralized, meaning that should an individual deviate from common practices, other individuals (both relatives and nonrelatives) will experience a sense of moral outrage and may inflict punishment on the individual via informal sanctions, or at least gossip.

As Chaldeans from the villages moved to Iraqi urban centers, on to Detroit, and out to the suburbs, the benefits derived from cooperation may have diminished, but many of the cultural beliefs, values, and practices supporting such cooperation would persist through inertia because they were so firmly interwoven into the fabric of Chaldean society. This could also explain why Chaldean families in Detroit were, and are, attracted to working together in family businesses, as well as becoming partners in other professions: the norm of working with family that developed when families worked their land together may have been superimposed onto their professions in the United States. In contrast, why Chaldeans tend to work in grocery stores and sell primarily tobacco, alcohol, and lottery tickets is likely an accident of history, as the first Chaldeans of Detroit happened to end up in this particular business. Though this is just one possible way that the high levels of cooperation among kin may have developed, it serves to illustrate that the unique cultural elements of Chaldean culture and history may have played a large role in shaping its history and the community as it exists today. Kin psychology and our evolved mechanisms for cultural learning combine to provide a larger set of insights into Chaldean life than either could provide alone.

When Cooperation Breaks Down

Although Chaldean families clearly cooperate abundantly, it would be misleading for us to portray the Chaldean family as an idyllic entity in which harmony and altruism flourish flawlessly. Evolutionary theory tells us that even close relatives should have conflicts. Individuals are related to themselves by $r = 1$, and to their siblings by about $r = 0.5$. This means that we should observe a wide range of circumstances in which individuals act in their own interest over those of their family. As in every human group throughout the world, conflicts do arise within families, and some people do take advantage of their own kin. In some cases, Chaldeans help their relatives greatly but resent doing so. Being part of a culture that places high demands on kin takes its toll, and at times people tire or anger at the constant assistance they are expected to provide to their relatives. Such was the case with the man who had 11 relatives living with him after they arrived from Iraq. He welcomed his family into his home, but after they moved out he chose to temporarily lessen his contact with the community so that he could have his own space and his own life without all the obligations that come with being part of the Chaldean community.

In more extreme circumstances, Chaldeans decline to help their relatives despite their kin psychology and the cultural prescriptions for aiding kin. These situations can become explosive, with permanent damage caused to the family. A woman told Natalie of her family arriving in Detroit and moving in with her paternal grandfather. Within a week, he told them to leave, and they moved to her maternal grandmother's house. Her family received no help from her father's relatives and now, 18 years later, they have almost no contact with that side of her family. This has influenced how she perceives the way that Chaldeans help their families in the United States. "It's really hard," she said. "Each person here looks out for himself. In Iraq, everyone helps each other." In another family, the father died shortly after arriving in Detroit, and the children relied heavily on help from their uncle (their father's brother). The uncle employed the children at his store, but he paid them $60 per week and paid his own son $80 per week. According to Hamilton's Rule (if we assume that dollars convert proportionally to fitness), the uncle should have paid his brother's children only $40 a week (and his son $80), assuming each worked at equal rates. The uncle was influenced by some factor that made him behave more prosocially than kinship alone would predict. We suspect that because there is a strong norm of helping kin, the uncle was afraid of third-party punishment (via damage to his reputation) and community gossip (chaps. 6, 7, and 8 explore these forces). Years later, one of the deceased man's sons bought a store with the uncle, but the uncle kept most of the profits plus interest from a loan he had given his nephew so that he could go in on the deal. Strong feelings of ill will persist. Their father had his own run-ins with his brothers during his lifetime, and he told his children before he died, "Never trust my brothers."

As in every group, some individuals do bad things to their family. Nonetheless, most Chaldeans live up to the cultural norm of helping relatives, and it is more common for relatives to both help and seek help from each other, even if it is done grudgingly.

Conclusion

Kin selection theory applied to human psychology predicts that most cooperation will be directed to people who have a high probability of sharing a gene for cooperation (i.e., people with a high degree of relatedness). Consequently, it is not surprising that Chaldeans tend to seek out relatives for help in most areas of their lives. Two aspects of this form of cooperation are noteworthy. First, even though Chaldeans appear to believe that they are all related and that they would help any member of their community because they are "one big family," people do not make the Big Mistake of confusing close kin with non- or distant kin. Rather, Chaldeans turn out to be very selective in choosing people with whom to engage in costly cooperative relationships, such as those that require the giving of money and time. In fact, different discounts for legal advice for close kin versus cousins by lawyers, and different payments to children versus nephews, shows that people are clearly calibrating the amount of cooperation or help they offer others by their degree of relatedness. Second, the exceptionally high levels of cooperation within families are likely a result of the particular cultural history of the group. It may have been the case that the economic and social structures of life in Iraq favored extensive family-based cooperation, and these norms have been carried over to the United States via cultural learning mechanisms and informal punishment that maintain robust forms of kin-based organization in this new environment.

The amount that relatives cooperate varies across cultures, as well as across domains within a cultural group, making it necessary to combine kinship psychology with theories of cultural transmission and cultural history to understand when, and to what extent, relatives help each other. Though natural selection acting on genes explains why kin cooperate, cultural learning theories are required for two reasons. First, cultural learning interacts with economic, social, and ecological factors to explain when relatives of a particular group help each other (in what domains), to what degree, and how they organize themselves. Norms about helping kin, including the threat of third-party punishment, may reinforce the strength of the behavioral pattern emerging from kin psychology. Second, cultural learning theory can provide an account of the temporal dynamics of change, helping us to understand why, how, and to what degree ideas, beliefs, values, and practices change over time.

6

Cooperation through Reciprocity and Reputation

> If someone helps you buy furniture, you aren't expected to give them money or furniture, but you will be expected to help them when they need help.
>
> —21-year-old female student

> Keep track of how much people give and then give them the same amount for their wedding.
>
> —28-year-old female, describing the reciprocation of wedding gifts

> It's all about gossip. People will be talking about who you asked for money, who gave and who didn't, and what kind of reputation you have.
>
> —Middle-aged woman describing the community's reaction to borrowing money

Though the interaction of culture, cultural evolution, and our evolved kin psychology can explain many aspects of human cooperation, a substantial amount of prosocial behavior remains to be explained. Looking beyond kinship, a variety of cooperative patterns in Chaldean life appear consistent with those expected from our theoretical discussion of direct and indirect reciprocity in chapter 3. This theoretical work suggests that Chaldeans should be particularly well suited to successful cooperation based on these two forms of reciprocity because Chaldeans tend to live in tight-knit, enduring, social networks: Chaldeans primarily live in heavily Chaldean-populated areas (such as Southfield), attend a Chaldean church, belong to Chaldean social clubs, attend many of the same weddings and funerals, and marry and socialize principally with other Chaldeans from within their community. There are multiple ties between members of the community and the same people interact with each other regularly throughout their lives. Chaldeans are aware (or at least believe) that these people will share their social and professional circles for a very long time. Few people break out of the close community, with hardly anyone going *away* to university (although many Chaldeans attend *local* universities), let alone moving to a more distant city or another state. People strongly expect that their current relationships will also be their future relationships. Taking this as a foundation, we should expect to observe a variety of cooperative patterns based on reciprocity.

However, because the real world of Chaldean social life mixes up and inter-twines the psychological effects of direct and indirect reciprocity, as well as many other human motivations, sharp laboratory-style tests are rarely possible. Thus, in line with one of our central messages about the integration of experi-mental and ethnographic approaches, we will first briefly review some key experimental findings on direct and indirect reciprocity, and then take a look at Chaldean life with our theoretical and experimental results in hand. The available laboratory data allow us to confirm some of the central predictions (those involving reputation and repeated interaction) derived from the theoreti-cal work. Then, adding living flesh to our theoretical skeleton, our ethnographic material shows reciprocity in operation 'on the ground,' and in a manner that is illuminated by the theory. To do this, we review some of the predictions and patterns discussed in chapter 3 about direct and indirect reciprocity, and then describe the ethnographic findings in a fairly general way, which still permits us to highlight the predicted patterns. This combination of experimental and ethnographic work allows us to study both how human psychology is geared to respond to reputation and repeated interaction and how cultural evolution has constructed institutions and belief systems—under the practical constraints of informational ambiguity and a limited memory—to take advantage of these evolved psychological mechanisms.

Experimental Findings on Reciprocity

Much of the experimental work on cooperation places people (usually universi-ty students) in tightly controlled game situations in which they must make decisions that will influence the amounts of money they and others receive. Empirical patterns that are markedly consistent with some of the most basic predictions of reciprocity theory have repeatedly emerged. Such results demon-strate that people respond strategically to (1) possibilities for future interaction between themselves and other players who are also making these monetary decisions, (2) the costs and benefits of cooperation, (3) reciprocity from others, and (4) reputation-building opportunities.

The Prisoner's Dilemma (PD) described in chapter 3 has frequently been implemented in the laboratory using real money. Recall that the PD involves two players who must make a decision about whether to cooperate or defect in an interaction. As captured in table 6.1, if both players cooperate, the pair jointly does best, each receiving $5. If one person cooperates and one defects, the defector does better (getting $6) than she would have if she had cooperated, while the cooperator does substantially worse (getting $0). Jointly, the pair does worse than if both had cooperated, receiving a combined total of $6 instead of

Table 6.1 Payoffs in Prisoner's Dilemma

	Cooperate	Defect
Cooperate	5,5	6,0
Defect	0,6	1,1

the $10 that would be obtained from mutual cooperation. If both players defect, both do better than the cooperator in a cooperator-defector pair, but worse than a cooperator in a pair of cooperators—each receiving $1 (better than $0), with the group netting $2. If the game is played only once (no repeated interactions), and the players are anonymous (no reputations), a purely self-interested individual who understands the game will always defect. Interestingly, however, when one-shot anonymous games were implemented among university students, and among small-scale farmers in rural Chile, about 50 percent of players cooperate and 50 percent defect.

To get at the effect of repeated interaction with the same player, we must compare one-shot games with repeated games. If human psychology is influenced by the evolutionary logic of direct reciprocity, then people should cooperate more when they expect to interact with the same person in the future. A simple formula allows us to calculate the conditions under which a tit-for-tat (TFT) individual should cooperate in a pairwise interaction when he believes the other individual is also a reciprocator. In chapter 3, we learned that cooperation can be sustained among two individuals if $\omega b > c$, where b and c are the benefits and costs in one cooperative interaction, and ω is the probability that the interaction will continue for another round. To make the relationship more intuitive, ω can be replaced with the expected number of future interactions, k.

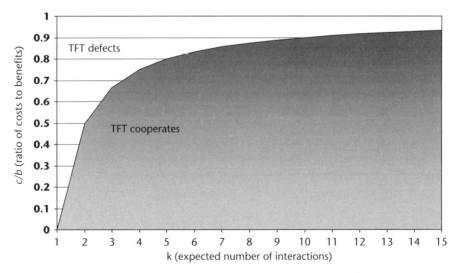

Figure 6.1 Regime of cooperation in a two-person interaction involving TFT.

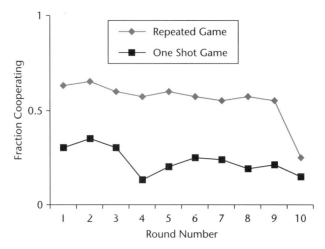

Figure 6.2 One-shot and repeated PD compared. Plot adapted from Cooper et al. 1996.

This yields (k–1/k)>c/b. Plotting this, figure 6.1 shows regions where reciprocal strategies, like TFT, should cooperate and where they should defect.

Do the insights from this theorizing pan out? Well, mostly. Results from Cooper et al. 1996, in which people played the Prisoner's Dilemma, provides a direct comparison of a series of one-shot interactions with a repeated game (partners were matched for ten rounds). Two aspects of these findings are noteworthy. First, and most notably for this chapter, people responded to the repeated game by cooperating more right from the first round (fig. 6.2). This confirms the qualitative prediction that repeated game incentives matter, and it is not merely trial-and-error learning; people see repeated opportunities for cooperation and respond with more cooperation. Second, cooperation drops off in the second-to-last round (round 9) to about the level that we observe in the one-shot game (but not to zero cooperation). The c/b ratio in this game is about 0.35 (extremely low), which means that only two rounds of interaction are necessary for cooperative play to be stable (see fig. 6.1). The theory also predicts that cooperation should drop off precisely in round 9, as it does, because there are no interactions after round 10, so there can be no penalty for defecting in that round. However, the amount of cooperation in both round 10 and in the one-shot game are significantly above zero, contrary to the theory's prediction. The implication of these findings is that direct reciprocity, as an evolutionary force, has shaped human motives, but that there is more to human social motives than reciprocity.

In chapter 3, we distinguished evolutionary theories built on repeated interaction from those based on reputation. In the context of indirect reciprocity, reputation involves information about a potential partner's history of past behavior that can influence an individual's decision about whether to cooperate or defect with such a partner. Theoretical work on indirect reciprocity suggests that reputational information used in one-shot interactions can maintain cooperation as long as the quality and integrity of the reputational information is robustly maintained and cooperative groups are quite small. Experimental work comparing one-shot Ultimatum Games with reputational Ultimatum Games shows the psychological effect of reputation-building possibilities (Fehr and Fischbacher 2003). In the Strategy Method Ultimatum Game (SMUG) used here, two anonymous players are allotted a sum of real money to divide. Player 1, the "proposer," must offer a portion of the total sum to Player 2, who is called the "responder." Player 2 must specify, before hearing the offer, the minimum offer she will accept. If Player 1's offer is equal to or greater than this minimum threshold, Player 2 receives the amount of the offer, and Player 1 receives the remainder. If Player 1's offer is below this threshold, both players get zero.

Figure 6.3 shows the results from two versions of the SMUG played over 20 rounds, from Fehr and Fischbacher (2003). In each round, proposers had 10 MU (MU is a monetary unit that converts to cash after the game) to divide with a responder. In one experimental version, players interacted in ten consecutive one-shot interactions (each round with a different person) without any knowledge of their partner's history of previous play; this was followed by ten rounds of one-shot games in which the full history of each partner's previous plays was known. The other experimental version was identical except that the ten rounds with reputational information preceded the no-reputation condition. It is important to notice that there are no repeated interactions in this setup, so players' behavior is not ascribable to direct reciprocity. Figure 6.3 shows the effects of reputation-building opportunities on the threshold of acceptance set by the responders over 20 rounds—a "threshold of acceptance" is the minimum offer that a responder is willing to accept.

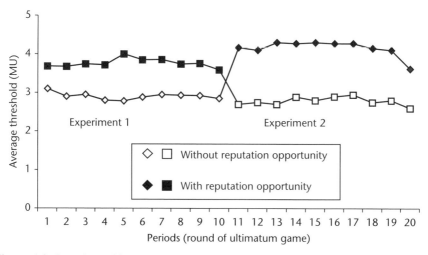

Figure 6.3 One-shot Ultimatum Games with and without reputation-building opportunities. Adapted from Fehr and Fischbacher 2003.

Theoretically, because the groups were small (two people), and the reputa-tional information was perfect, reputation-based theories predict that indivi-duals in this situation should be keenly sensitive to reputations and reputation building: setting a higher threshold early in a game with reputational informa-tion will reap benefits in later rounds. Consistent with this, as soon as reputa-tion-building opportunities are added to the game (experiment 1 in fig. 6.3), responders jack up their acceptance thresholds (and thereby force proposers to give more of the total pie). Because no learning was required, and experience has little effect, it seems that people are psychologically primed to respond to reputation-building opportunities and readily adjust their behavior. Along these lines, in the reputation condition, players drop their thresholds in the last round of the reputation treatments (round 10 in experiment 2 and round 20 in experiment 1), presumably because there are no future opportunities to reap reputation benefits.

Beyond these basic findings, we should also highlight: (1) players in the no-reputation portions of experiments 1 and 2 set their minimum threshold much higher than can be explained by any reciprocity-based theory; and (2) experi-ence did not drive down responders' thresholds in either treatment. Again, although reputation is clearly important and operates as predicted, there is more to human social motives than merely direct reciprocity, reputation, and kinship (which are entirely ruled out in the first half of experiment 1 and the second half of experiment 2).

Recent work (Engelmann and Fischbacher 2002; Seinen and Schram forth-coming) has directly addressed the effects of indirect reciprocity on human motives and decision making. Engelmann and Fischbacher's experimental design elegantly disentangles two motivations for helping: 1) building up one's own reputation as a "helper" so that one will be more likely to be helped in the future, and 2) helping people because they helped others in the past. Sixteen subjects played 80 rounds of a "helping game." In each round, players were randomly matched and randomly assigned the roles of donor and recipient. Donors had to choose between doing nothing and paying 6 to give 15 to their recipient. For half of the rounds (either the first 40 or the last 40), player decisions were incorporated into their donor score, which was provided to future players who played the role of donor for that individual. Donor scores were simply the number of times out of the last five opportunities for donation that the individual donated. For the other 40 rounds, the individual had no score. This design allows us to partition helping behaviors conditional on the recipient's donor score and the donor's ability to improve his own donor score.

Figure 6.4 summarizes two key findings. First, regardless of whether their donation can affect their own reputation, individuals donate more to others who themselves have donated more in the past. However, the level of donation increases when donors can enhance their own reputation by helping others, especially when recipients don't have all that great a reputation themselves. To see this final point, compare the difference between the helping of "donors with scores" to helping "donors without scores" at the recipient's scores of 0.40 and 1. At 0.4, the helping rate increases by 600 percent when donors can improve

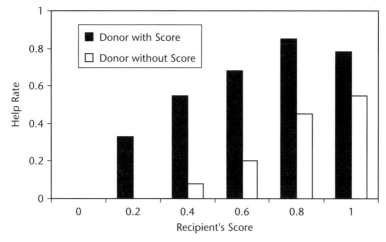

Figure 6.4 This figure shows how the rate of helping varies according to the potential recipient's helping score, and according to whether the donor can affect his reputation. Adapted from Engelmann and Fischbacher 2002.

their own reputations ("donors with scores") compared to when donors can't improve their own reputations ("donors without scores"), whereas at a recipient's score of 1, the helping rate increases by only 50 percent. Moreover, even when the recipient helping score drops as low as 0.2, more than 30 percent of donors with scores will help, whereas no donors without scores are willing to help.

This finding is consistent with theoretical predictions arising from the indirect reciprocity models discussed in chapter 3 (e.g., Brandt and Sigmund 2004; Ohtsuki and Iwasa 2004; Panchanathan and Boyd 2003: 123). Successful strategies (1) should always consider the good reputations of others in deciding to help them, and (2) can often consider improving their own reputation by helping others regardless of the recipient's reputation. This work shows that students consider both their own reputation and that of their assigned recipient. When donors strategically help to increase their own scores, indirect reciprocity models overlap with the signaling models (Bliege Bird, Smith, and Bird 2001) discussed in chapter 3. As we'll show below, using generosity or helping behavior to elevate one's reputation (i.e., one's donor score) is an important element in understanding Chaldean cooperative behavior.

Using an experimental design similar to that of Engelmann and Fischbacher,[1] Seinen and Schram's work corroborates the finding that European students help more when the person they are helping has helped others in the past. It also confirms two additional aspects of our theory: (1) students are sensitive to the ratio of costs to benefits (c/b), such that they help less when c/b is greater, and (2) over many rounds, different experimental groups evolve different norms about how much helping behavior is sufficient to render a player "worthy" of receiving help. Both of these finding are consistent with our broad theoretical expectations, but the emergence of norms will be dealt with in the next chapter.

Theoretical Expectations in Ethnographic Analysis

Direct Reciprocity

Using the body of theoretical work on direct reciprocity discussed earlier, several interesting patterns emerged from our ethnographic inquiry. To remind the reader of our discussion in chapter 3, we list here the key predictions or patterns that are relevant for the ethnographic results below.

1. Direct reciprocity is unlikely to favor cooperation when cooperative groups are large. Dyadic forms of direct reciprocity are the ones most likely to be sustained in the long run. This suggests that long-enduring institutions, which are based on direct reciprocity, will involve cultural practices that transform public goods dilemmas into situations of dyadic cooperation. Below, we show how Chaldean practices at weddings and funerals transform what would otherwise be a public goods dilemma into a series of discrete, long-running, dyadic interactions.

2. The expected patterns of reciprocity can be partitioned into two categories. These two categories allow individuals to make effective use of the benefits of reciprocal cooperation, while dealing with (1) memory limitations (for a large number of partners), (2) errors, mistakes, and/or noise in one's assessments of the contributions of others, and (3) differences in the expected number of repeated interactions for different people.

   a. The first category involves a small number of "exchange partners" (friends). Exchanges are numerous and across many different domains (rides to the airport, child care, etc.); as a consequence, the probability for errors in any given cooperative contribution is high, and thus (in accordance with the theory) people are generous, forgiving, and contrite.

   b. The second domain involves a substantially larger number of partners. Interactions are rarer and are very trackable (often 1-for-1 exchanges, in-kind); written records are often kept, if possible, but what matters is the most recent transaction. Because errors are not likely, reciprocal returns are either exactly equal or slightly generous, and partners are highly provokable.

3. Culture provides standards or guidelines for how generous one should be when information is not available from previous interactions. For example, social groups will have culturally transmitted prescriptions about how much to give on the first interaction. Reciprocity theory (and thus our reciprocity psychology) provides only constraints (maximum c/b value for cooperation); it cannot tell what to give, or how much the other will think is right.

4. Playing off our kinship psychology, culturally transmitted beliefs can modify the expected time horizon (k) for our reciprocity psychology and thus strengthen the psychological effect of the "shadow of the future" (i.e., the effect of anticipating future interactions). If reciprocity occurs among families, and the actions of members of one generation are taken as cues

about the strategies that will be pursued by subsequent generations of that family, then the action of a father will affect not only the behavior of another family toward himself, but also the behavior of that family's descendants to his children and grandchildren.

Indirect Reciprocity

Chaldean social networks are also well suited to cooperation via reputation-based indirect reciprocity. First, as noted above, Chaldeans maintain a close-knit group that leads to expected (and realized) long-term, repeated interactions with the other members of the community, which places a great emphasis on maintaining a good reputation. Consequently, the potential benefits from being a good reciprocator are very high, as are the costs of being punished by a damaged reputation. Second, Chaldeans love to gossip, and people say that they either know what is going on with everyone in the community or can find out with just a couple of phone calls. This taste for gossip, combined with numerous overlapping and redundant pathways among individuals and little migration out of the community, creates dense information networks and an environment in which reputation is of the utmost importance. The closeness of the community makes it very difficult to hide any offenses that could damage one's reputation, and news of a transgression spreads rapidly. By the same token, good behavior that can improve one's reputation can also spread quickly through the group and create long-term benefits. As we mentioned in chapter 3, however, a theoretical problem remains in explaining the maintenance of the practice of spreading high-quality information and punishing those who do not (and those who might spread false or low-quality information). We explain how this is solved in chapter 7 by describing how indirect reciprocity is strengthened and reinforced by social norms that prescribe the spreading of truthful gossip and the punishing of those who don't share reputational information, or who spread false information.

Several of the theoretical deductions made in chapter 3 are developed here. These are:

1. When enmeshed in tight social networks, people will be more concerned about reputation.
2. When situations of high "broadcast value" arise, capable individuals will perform costly acts of help in public settings in which many people may observe their cooperativeness, regardless of the perceived helpfulness of those being aided. This allows individuals to make the most of the benefits of having a good reputation with showy public displays.
3. Generalizing from #4 in the section above, cultural beliefs that link an individual's behavior (or reputation) to her close kin can magnify the importance of reputation for maintaining cooperation based on indirect reciprocity. For example, if a father develops a reputation for nonhelping behavior, his sons may inherit that reputation and have a difficult time shaking it.

4. In ambiguous situations, individuals will have an incentive to fake helping behaviors. That is, when it is difficult to tell whether an individual is actually helping, people will attempt to signal that they are helping when they actually are not.
5. Individuals will use ethnicity as a cue of participation in the same social network. This will lead to more cooperative, trusting behavior toward coethnics and to less cooperative, trusting behavior toward people outside the ethnic group.

Reciprocity "on the Ground"

Contributions to Funerals

Chaldean cultural practices take advantage of their tendency for long-term interactions by transforming potential public goods problems into a series of dyadic reciprocal interactions. To illustrate this, consider the Chaldean funeral tradition of giving money to the family of the deceased to offset funeral expenses. At the funeral home, a box is set out to collect donations from friends and relatives. One possible setup would be for everyone to leave money anonymously (by putting in cash, for example), and at the end of the funeral the family opens the box and finds out the collective total of all the money given by the mourners. This would fully satisfy the purpose of the donations: to get money to the family of the deceased to use for the funeral expenses. Giving the money in this format would create a public goods situation in which the family of the deceased would feel a generalized gratitude toward all of the people who came to the funeral, including those who may have slipped into the funeral without giving any money at all. Free-riding would be easy, and the family would have no way of ascertaining who gave a donation or how much. In this scenario, the tradition of leaving money could quickly unravel. Some people, even if only a few, could leave little or no money, thus driving down the average contribution per attendee. The family of the deceased, considering these diminished amounts, will not want to give substantially more at future funerals than they received, and so they may give less than their previous donation at the next funeral they attend. Over time (perhaps over generations), the amount donated would gradually decline to near zero. This effect has been repeatedly shown in experimental work on public goods games (e.g., Fehr and Gächter 2000, 2002); some of this work is described in chapter 7.

Instead of this format, however, Chaldeans put their funeral donations in an envelope marked with their name. When the family opens the box after the funeral, they know not only the total amount of money they received but also who gave and how much. As explained in many interviews, the family records the amount given by each person and uses this as both an indication of how much the giver cares about the deceased and his family, and as a guide for how much to give when this family has a funeral. By using this method of donations, a dyadic relationship is created between the recipient(s) and each giver rather than the public goods problem described above. This allows a recipient to directly reward a generous giver by delivering him a large donation when someone in his family dies, and an indirect reward can be given by telling people about the substantial donation—thus improving the giver's reputation as a caring and generous person (more on this reputation signaling below). Structuring exchanges in this way enables individuals to acquire high-quality, reliable information about their partners—which they can use to decide whether to pursue future interactions—and to shape the person's reputation by sharing the information with other people.[2]

Funeral donations come from a wide circle of distant kin and family friends. Consistent with our theoretical discussion of reciprocity, Chaldeans play a fairly exact game of tit-for-tat with funeral donations: they try to give back whatever they were given. An elderly woman said that she had recently gone to a man's funeral and "gave an envelope" (donated) "*because* he gave one at my husband's funeral" (emphasis added). Jennifer, a 28-year-old who works in the entertainment industry, explained that when her father died, her family kept track of who gave them money and how much they gave. When they go to funerals, they check their records and give the same amount that they received from that person. She cited in particular one man who gave her family $30 when her dad died, so she gave him $30 when one of his parents died five years later. This sort of description was common among our informants. Such examples illustrate the importance of culturally evolved technologies (writing and the practice of recordkeeping) in maintaining effective cooperation over the long term. This woman knew exactly how much the man had given at her father's funeral, and she knew exactly how generous she was going to be with him. If someone free-rides (partially or fully), he can't ride it out and hope that eventually it will be forgotten: Chaldeans keep track of levels of cooperation from other Chaldeans for a very long time.

A few people explained that the amount that someone gives you at a funeral is a cue of the minimum you can give to them. For example, Tom, a 40-year-old insurance broker who came to Detroit at age 18 by way of Greece, said that you "keep track of the amount that people give, but you don't give the same amount [to them] because it can be many years later, and things have gone up." Tom seems to be playing "inflation-adjusted tit-for-tat." As explained in chapter 3, such strategies should be expected, as generous strategies can coexist successfully with TFT strategies and make the whole population more resistant to the destructive effects of noise. In this environment, individuals keep written records of donations, but inflation acts as a kind of noise that would gradually bring down the value of donations. Generous strategies, even at low frequencies, can prevent the overall levels of cooperation from being driven down by forces such as inflation.

Consistent with the expected theoretical patterns, these funeral exchanges (1) involve a large number of people (most of whom are not close friends), (2) are tracked exactly and kept "within domain" (in deciding what to give back, you don't, for example, consider exchanges of babysitting or driving), (3) are either exact or somewhat generous (no one reported a rule of giving back a bit less than what he got; in the long run, such a strategy would drive cooperation to zero).

However, as one would expect in real life, there is much more going on than just direct reciprocity. Donations that are particularly high or low relative to either "the norm" or to previous exchanges between the pair have more general effects on that person's reputation. Our interviews indicate that many Chaldeans see the size of a funeral donation as revealing information not only about the donor's generosity but also about other of his qualities, so multiple aspects of a person's reputation are affected. For example, Rose explained that the size of the donation lets you know how much the person cared about the deceased and how concerned they are about the family. Another woman said that the amount that a person gives reflects how much he respects you. Since traits other than generosity are associated with funeral donations, it is easy to damage a person's reputation if he gave little or nothing. By telling other Chaldeans that someone gave a small donation, one is saying that the person is cheap, lacking in respect, not a "good Chaldean," and short on caring. This suggests that the potential costs of reputation-damaging gossip may be high because people will not want to form relationships with people who have these traits. Very little theoretical work has addressed the issue of how cultural evolution might shape the contents and extent of reputations, although Panchanathan and Boyd (2004) have opened up the issue.

When someone gives a large donation, they may be looking to improve their reputation or raise their potential returns. Lena, a 65-year-old homemaker who has lived in America for 35 years said that her rule is to "check how much the person gave you, and give more than they gave." Rather than adjusting for inflation, Lena seemed concerned with ensuring her reputation. Emphasizing the effect of good behavior across domains, Karen explained that a person is more "likely to help someone more if they gave you a lot."

Weddings

Showing many of the same theoretical patterns revealed at funerals, weddings are another situation in which Chaldeans form reciprocal dyads and play tit-for-tat-like strategies with gift giving. In the typical Chaldean pattern, the groom's parents are responsible for paying for the wedding, including jewelry and clothes (though the bride's family pays for the engagement party), and the bride and groom receive all their gifts in the form of cash, which is given to the parents to cover the wedding costs. Because nearly all the services (e.g., photographer, music, food, hall) are provided by Chaldeans, who are well acquainted with the cash-gift tradition, many of the wedding bills are paid after the wedding, when cash is available to cover the expenses. With this gift structure, the newlyweds and their parents know the exact amount of cash given by each guest. When a married couple subsequently attends a wedding of one of their guests, the amount that the guest gave determines their own gift. Jane, for example, explained that "[You] keep track of how much people give and then give them the same amount for their wedding; it is fair and respectful to give back the same amount." As with the funeral donations, the amount of the gift is strongly influenced by tit-for-tat reciprocity (it's "fair and respectful" to give the same amount).

The tradition of using cash in labeled envelopes eliminates any ambiguity in the value of the gift. First, it avoids the public goods problem that would be created if everyone just threw unlabeled wads of cash into a box by transforming it into a large number of dyads. Second, it avoids the ambiguity ("noise") in valuing gifts such as blenders, plates, dishes, wine, and luggage. As theory has shown us, increased ambiguity would call for different—generous and forgiving—strategies to maintain long-term stable cooperation. Labeled envelopes filled with cash maximizes the success of TFT-like (i.e., equal-return and highly provokable) strategies, which negates any need for tracking the history of each relationship.

Like funerals, Chaldean weddings also illustrate how behavior extends to one's close kin. Michael expressed this, along with several other noteworthy sentiments, when he stated that "a wedding is like a debt; someone comes to your wedding and gives money, then later you go to their wedding and give money. . . . After the wedding, parents keep track of how much each person gives and then gives the same amount to them later." Thus the debt extends not only to the couple that got married, but also to their families.

The same generosity evident in some donations to the funeral box is also observed in wedding gifts. "Also like the funeral, there are some people who give a little more than they received, rather than exactly matching the gift," Michael said. Rita, a 31-year-old woman who has been married for 10 years, is one of the people who prefer to "err" on the side of generosity: "I kept track of who gave money and how much. Then, when I go to their wedding, I give a little more than they gave me."[3]

In addition to equal or slightly generous reciprocity at weddings, people also explained what to do if one lacked information on past exchanges with the families hosting the wedding. Everyone interviewed agreed that each guest should give $50 ($100 for a couple). The $50 amount provides a reliable, widely accepted rule of thumb that individuals can draw on if they lack a past history of exchanges with the relevant families. Setting an accepted standard also provides a mark by which to measure generosity and stinginess. Notably, people were also quick to specify that this rule of thumb does not apply if you are a close friend or relative, in which case you give more.[4] Consistent with our theory in chapter 3, such statements suggest that people partition exchange partners into three broad categories—relatives, close friends, and other reciprocal partners ("distant friends")—that are subject to different rules.

It is worth reminding the reader that the rule of "everyone gives $50" is fair as long as that is what people have generally accepted as fair in a community. There is no absolute standard from which to judge fairness. This rule, for example, does not take into account the fact that people have different incomes, different amounts of wealth, different allocations of talents, and different num- bers of children. The "cost" of $50 to someone who is relatively poor could be quite high, whereas $50 may be an insignificant sum to someone who is wealthy. An immense number of different allocation rules could be made up, all of which would be equitable or justifiable from some human perspective, but the key aspect of any rule that makes it fair is that most people agree on it as the rule.

Chaldeans also use reciprocity in domains in which it is nearly impossible to use cash as the currency of exchange, as when deciding who gets invited to a wedding. As it was told to Natalie, "If you were having a wedding, you'd invite your immediate family, then more distant family, then friends, and then *people who had invited you to a previous event*" (emphasis added). Once again, we see how Chaldeans partition the social world: relatives of differing closeness, friends, and other exchange partners, with tit-for-tat exchanges tracked closely for the largest circle of exchange partners and measured in-kind within a domain (e.g., wedding gifts, wedding invitations, and funeral donations). As we explained in chapter 2, this is not because things such as money, physical efforts, and invitations cannot mentally be converted into a single currency. Rather, it is because such conversions introduce noise, and noise tears apart reciprocity in these kinds of situations.

Within a smaller circle of people, Chaldeans also follow the rules of reciprocity, although they are somewhat different. Even when it comes to family and day-to-day help, people follow rules of reciprocity; within tight-kin groups, reciprocity and kinship can combine to achieve higher degrees of cooperation than either could alone. Rita described how she helps her sister by driving her children to school, and then she added that her sister reciprocates the help. Her reason for adding this was not clear, although she may have been letting us know (signaling) that she doesn't let people take advantage of her and that she helps people as long as they help her. Rose, a single mother with some financial pressures, said that her friends and family help her out by bringing her food, helping with household repairs, and giving her money. Like Rita, she added (without being asked) that she returns the help; she isn't comfortable giving cash to the people who help her because she doesn't have enough money to give them an amount that she feels is adequate, but she does buy them things she can afford and helps them in other ways. Rita appeared to be announcing that she doesn't free-ride and that she is a good person with whom to form a reciprocal relationship. As expected, this kind of close-knit reciprocity is not confined to a single domain (people exchange all kinds of things), involves exchanges that are difficult to track and are inexact, and doesn't extend to a wider social circle (e.g., Rita and Rose don't have these reciprocal relationships with everyone who attended their weddings).

One final note of interest on Chaldean weddings: in a pattern consistent with our discussion in chapter 4 of the evolution of investment in children and the movement away from the communal model of households, some of the details of who pays for the wedding and where the gift money goes is beginning to change. Three people explained that parents will now sometimes pay for the wedding and not get paid back. Specifically, one man said, "Some very wealthy parents may pay for the wedding and buy them [the bride and groom] a house and a car and not expect to get paid back!" This is not the typical Chaldean pattern, but it hints at cultural changes in the direction we'd expect from the research presented in the last chapter.

Reputations Matter, or at Least Chaldeans Think They Do

For cooperation to be enhanced by reputation, reputation has to matter, or at least (and more important), people have to believe that reputation matters. Here we show ethnographic data demonstrating that people believe that the benefits associated with a good reputation are real, and that every Chaldean should care about his reputation. Some of the benefits are easy to identify, whereas others remain fuzzy. For instance, within the Chaldean community, a person with a good reputation is more desirable as a marriage partner, more likely to be hired, more likely to be lent money or extended credit, and more respected. For example, as you'll see in chapter 9, Robert credits his good reputation in the community for allowing him to acquire many clients and enabling him to advance his career in financial advising. A good reputation also helps in the retail/wholesale industry. A wholesaler said that he chooses his Chaldean customers (the grocers that he supplies) based on the person's reputation or the reputation of the person's family. These people are given credit immediately, and no contracts are signed:

> Everyone knows which families are good or bad, and you just do business with the people who come from good families—I go by the family name. If I don't know someone, I call people and ask about his family. People just mention who they are and [if they have a good reputation], they get credit.

The fact that Chaldeans interact so extensively and have a fondness for gossip means that the wholesaler need only make a few calls to determine if a person will likely be trustworthy in a business relationship. This method does not apply to his non-Chaldean clients because he lacks a mechanism for learning about their reputations. As a result, non-Chaldean clients sign contracts and have to give postdated checks for their first few orders before the wholesaler will extend them credit. A Chaldean with a good reputation who is well integrated into the community (and hence is known by reputation) has an easier time forming profitable business relationships with other members of his group.

Outside of business, the benefits of a good reputation are equally apparent. When people choose someone to date or marry, they consider the reputation of the individual and his or her family. However, Chaldeans seem to be more concerned with avoiding involvement with someone who has a bad reputation than in seeking out a partner with a good reputation. The opinion that people have of an individual also affects the likelihood that he will be lent money. Many explained that a person can ask family and close friends for a loan to start a business or purchase a home. Usually, the person who is asked will lend the money (if he has it) and does not require any contract or, normally, interest. But if the person asking for the loan lacks a good reputation, then the person will not lend him money or, at best, will require a formal contract. As shown in the experiments described above, the benefits that accrue to a person with a good reputation create an incentive to do good things even when these actions are costly.

Costs of a Bad Reputation

The flip side to the benefits of a good reputation is the punishment, and consequent costs, dealt to people with bad reputations. Among Chaldeans, a person with a bad reputation is less likely to be given credit, to be hired, to be desired as a business partner, or to be lent money. People will also not want to date or marry a person with a bad reputation, and if they do so, people will gossip about them. Zev, an American-born, 53-year-old speech pathologist, gossiped with Natalie about a man who had recently gotten divorced. He said that everyone knew that the man should never have married the woman because "she has bad parents, [and] the apple doesn't fall far from the tree." He then told Natalie about a man who is "dating a girl who seems very nice, but he shouldn't go out with her because she has a bad family."

This brings up an important component of reputation in the Chaldean community: a person's actions reflect upon his whole family, and, conversely, the reputation of a family applies to all of its members. People recognize that a social transgression will have effects on both them and their families. Jennifer has dated non-Chaldean white men but said she knows that she can never date a black man, because if she did, no Chaldeans would let their son marry her, and she would disgrace her family. Similarly, Rose said that her divorce caused her to be shunned and her family to be disgraced. Joseph said that his divorced brother now has a reputation problem in the community, and that the divorce has hurt his parents' reputation and caused them shame. Because the reputation of an individual attaches to his whole family, a person who (on his own) would be willing to incur the cost of a bad reputation for himself might reconsider, as his actions would create costs for his entire family. Here, reputation interacts with kinship to compound the costs of punishment and to create even greater incentives to act in a way that the larger group considers appropriate.

As explained in our discussion of the Panchanathan and Boyd model (2004) in chapter 3, cooperation maintained through reputation (indirect reciprocity) can also maintain other costly norms—which might have nothing to do with cooperation—once reputation is generalized beyond what happens in the indirect-reciprocity cooperative encounters. Among Chaldeans, punishment for violating group norms, such as prescriptions about whom to marry (a norm unrelated to cooperation), frequently includes gossip that damages the transgressor's and his family's reputation. The immediate effect is identified as a "loss of respect" or being viewed as "selfish," "unloving," or, generally, not a good person. In the longer run, damage to a family's reputation appears to have more tangible consequences, as family members are refused loans or get turned down for dates—just as, in the Panchanathan and Boyd model, reputation damage from some costly norm violation causes losses (punishment) through the "helping interaction." The negative effect on one's kin creates a situation in which a person is often better off doing costly (perhaps undesirable) deeds based on group norms rather than suffer the even greater cost of having the whole family punished.

Manipulation of Reputations: Costly Signaling and/or Fake Signaling

Boasting and Spreading Good Reputational Information about Yourself

Indirect reciprocity and reputation create incentives for people to spread "good gossip" about themselves. Coupled with their close-knit community, when someone does something, good or bad, a large segment of the community knows within a relatively short period of time. Chaldeans often help to fuel the spread of information about their good qualities or deeds by openly telling others about them. People frequently described their accomplishments or good traits, as when Sue informed Natalie that her family is "well respected in the Chaldean community" and that she was "taught honesty, integrity, family, and sincerity." An unemployed woman in her thirties noted that she was "very good in school, very smart. Everyone said I should be a doctor." And a young man who recently started his own business told Natalie that the company was doing well and growing fast, and that he had much more money that month than when she spoke with him the previous month. There is a strong incentive to make sure that people know about your good qualities and acts because a person reaps rewards when he does things that the group thinks are good. Although many societies in the world have culturally evolved beliefs and practices (norms) that restrict individuals from spreading such pro-self information, the Chaldeans are not one of these groups.

People in the Chaldean community recognize which behaviors, ideas, and values are perceived as "good" (reputation benefiting) by their group, though they may not agree with them or even be consciously aware of them. One favored behavior/value is helping others, especially when it is directed toward helping Chaldeans, both in metro Detroit and Iraq. This creates a motivation for people to participate in cooperative activities, as it will improve their reputation. Again, we stress that this motivation need not be conscious.

If reputational benefits are a key component as to why Chaldeans cooperate, then we should expect to see people preferentially engaging in highly visible forms of cooperation. This is, in fact, what we found happening frequently in the Chaldean community. When people act charitably, either by giving their time and labor or by giving money, they often do so in a way that others will know about it. This public display of pro-group behavior is consistent with signaling theory. Several people told Natalie that they prefer to give charity in a form that will let others know that they were the ones who gave it. For example, Sue doesn't like to put her cash directly into the collection at church, preferring to put it in an envelope marked with her name so that the church will know how much she has given. And Lida, an active church member, remarked that the number of people using envelopes is increasing. "At church collection, people can give anonymously or use envelopes that you put your name on," she said. "Ten years ago, about 10 percent of the people used envelopes. Now, about 35 percent use envelopes. They use the envelopes because they want you to know how much they give."

A young man also described his preference for making donations publicly. Natalie had asked him a hypothetical question about a wealthy man who didn't contribute to a church collection that was raising money for Chaldeans in Iraq (this study is detailed in the next chapter). As part of his reply, he said that if the man is wealthy, then he would probably give a large sum for this cause, in which case he wouldn't give it in the anonymous collection. Rather, he would probably give the donation at another time when he could donate it in front of the congregation and have his generosity observed. "[He] would want everyone to see and know what he gives," he said. He added that when he has more money, he will make his donations publicly.

Chaldeans value generosity (in certain situations), and this may lead people to act generously as a way of signaling their value as a cooperator and as a good person who embodies the group's norm of generosity. This signaling may be at play when Chaldeans fight over who gets to pay the bill when eating out. A 26-year-old male grocer explained:

> We're very generous. When a bill comes at a restaurant, everyone wants to pay the whole thing. Everyone is generous and doesn't want anyone else to pay. This is true with friends, cousins, even *people you don't know* [emphasis added]. When youths are 16 or 17, they all chip in to pay the bill, but people in their twenties all fight over the bill.

Note that this form of help is not based on exchange with long-term partners, as paying the bill applies even to people one doesn't know. Consistent with both our theory and the experimental data described above, "donors" can ratchet up their reputation with generosity regardless of whether they are helping other "helpers."

Generosity extends into other bill-paying domains, such as paying for each other's purchases. Mary told me that when she goes shopping with her sister, Nadine, they both try to pay. At the checkout, both women will thrust their credit cards at the cashier, trying to get the cashier to put the total of both women's purchases on her own card. Chaldeans take this generosity very seriously, truly wanting to incur the financial cost and to show that they are a good person by doing so. Mary described an upsetting event in which she had gone shopping with Nadine for peaches. Mary had bought several baskets of fruit, and Nadine bought only a single basket. Mary insisted on paying for Nadine's peaches, but Nadine got so angry over not being allowed to pay that she refused to take her fruit and wouldn't speak to her sister.

Although generosity in particular situations has become so internalized that Chaldeans may simply want to be generous because they see it as the right thing to do, their generosity may (without contradiction) be signaling that they are a good Chaldean with good Chaldean values. This may also be the case when people are generous with guests. Visitors are welcomed into the home and offered lots of food and drink. People explain that this generous hospitality is an important part of being a good Chaldean. Generosity, in appropriate contexts, seems to signal that the person is a good group member who exemplifies the norms and beliefs of the group (thus, this may operate as a cue of ethnic identity, as in chapter 9).

It is possible that generous behavior could be explained with direct reciprocity and that people take turns, over the long run, paying for each other's meals and purchases. However, using payment of restaurant bills as a model, we do not think that direct reciprocity applies here for four reasons. First, people fight over who gets to pay, rather than saying, "I'll pay this time, you pay next time." This makes it seem that people want to pay and incur the cost rather than to pay with the expectation of someone else paying next time. Second, when people eat out in groups, the composition of the group varies—the group eating together is not always the same. This makes it extremely difficult to use reciprocity and to keep track of who has paid for whom in the past. Third, when men and women dine together, the men always pay. This means that there is no possibility of reciprocation by the women, and yet the men continue to pay for them. Fourth, people pay for each other's meals even when they do not expect to have a future interaction. On numerous occasions, Chaldeans paid for Natalie when she was doing interviews in restaurants or cafés even though there was no expectation that they would dine together again. These generous acts stand in marked contrast to the typical reciprocal relationships observed in the community in which very accurate records of exchanges are kept and the favors are balanced. In these generous interactions, people seem to employ the strategy of shouldering a disproportionately great share of a cost as a way of signaling that they have the qualities of a good Chaldean.

Cooperation motivated by a desire for recognition or reputational enhancement can be inefficient because many people cooperate only as much as is necessary to get recognized as a big cooperator. It also creates an environment in which people try to fake cooperation cues without actually doing any work. A primary form of perceived cooperation involves participation in Chaldean organizations, in which it is very easy to free-ride. Within an organization, members can assess who really puts in time and effort to help the causes of the group, but outsiders can't determine which of the people who call themselves members actually contribute and which merely attend meetings and tout their affiliation with the organization. Karen belongs to a Chaldean organization and explained to Natalie why she doesn't feel that she needs to help with any of the group's activities in order to be a good group member. She claimed that she benefits the group by spreading word about the organization throughout the community, telling people about the group and what it does. She managed to perceive her own lack of effort as altruistic, whereas it seemed to us that she was free-riding off the group's reputation. Though she didn't help with any of the group's projects, she made certain that a lot of people knew that she was associated with this wonderful group, and thus her own reputation was enhanced.

Punishment of free-riders is difficult in this situation because they are not easily detected by outsiders, and within the group there are few ways to punish nonactive members other than chastising them (publicly or privately) or kicking them out. In the CARE group (Chaldean Americans Reaching and Encouraging—a charitable group of Chaldean teenagers and young adults), the organization's president periodically scolds the group as a whole for its overall lack of participation and tells people that it is unfair to make a few members do all the work. He tells people that they need to sign up for projects and carry through with their commitment. As best we can tell, this accomplishes nothing. It seems that without direct punishment for free-riding, such as explicitly mentioning the names of people who are insufficiently active or kicking individuals out of the group, participation levels are unlikely to increase. The general comments about member inactivity do not damage any individual's reputation, so there is little incentive for members to increase their participation, and yet the executive committee does not take the more aggressive step of kicking out slackers. Either the group feels that the presence of the nonactive members makes a positive contribution (perhaps Karen is correct), or the fairness violation doesn't merit the conflict that may be created by the harsher punishment. Nonetheless, it seems unfair (to many Chaldeans) that inactive members get recognition for being part of this group without actually doing anything.

Many Chaldeans recognize that there are people who participate in volunteer organizations out of self-interest. A leader of an organization said, "People join the groups because they want recognition from the community, not because they care about helping the causes." He added that in another Chaldean organization, little is accomplished in meetings because each member is busy telling the others how much he has done. Natalie attended one of these meetings and found his remark to be extremely accurate. During the meeting, the men self-aggrandized, repeatedly calling out their accomplishments and making certain that they would get credit from the group for everything they had done. A man who had recently made a speech to the government on behalf of the group repeated numerous times that everyone applauded and thought his speech was great. Later in the meeting, the president of the organization referred to a picture of one of the local Chaldean churches that had appeared in the newspaper. Immediately, another man began to call out that he and the man next to him were responsible for the picture's appearance, repeating the refrain several times.

Although the self-aggrandizers are helping their group and/or community, their reputations can be damaged if people think that their reason for joining the volunteer group was motivated purely by self-interest. Sue, for example, spoke harshly of a man involved in several organizations but who many believe has suspect motives. "He does things for personal advancement rather than the community," she said angrily. The difficulty in gaining a good reputation via signaling is that the actor must appear to be motivated by prosocial goals rather than self-advancement. Though prosocially motivated cooperation can be faked, the fakers may be found out if they overadvertise their good deeds.

Despite the free-riding in these organizations, we believe that more cooperation within the Chaldean community occurs because of the drive for positive recognition than if this motivation were absent. We make this claim for several reasons. First, some people who seek recognition (which may be an unconscious desire) also are committed to the causes of their organizations and accomplish tremendous amounts of good. Second, not all free-riders are total free-riders. In other words, these people may not do a lot of work within the group, but they do make some contribution of time, money, and effort. Third, even people who don't take on any active role in any of the projects or endeavors of the organizations still need to attend at least some meetings to claim member status. Their attendance may benefit the organization if they give valuable input during the meetings. If there were no possible reputational gains then these people might not participate at all.

Reputational Concerns Lead to Conflict

The desire for recognition from helping the community leads to conflict and competition between the various volunteer groups. All of the Chaldean organizations are working toward a common goal: helping Chaldeans. Yet, the groups each work independently so that they can get the full recognition for any accomplishments. Natalie was at a meeting of one of the dominant groups in which people were heatedly discussing what to do about land in Chaldean Town that the government was going to sell at a much discounted rate in order to bolster development in the area. Apparently, a prominent Chaldean man had given the governor a business card from a rival organization (the other organization also helps Arabs and it is charged that they claim to help Chaldeans in order to get funding but they don't actually carry through with this assistance). The people at the meeting were outraged that the other group may get to develop the land instead of them, and reiterated among themselves that they are the official spokespeople for the Chaldeans. It was decided they would call the man who had passed on the business card to remind him that their group represents the Chaldeans. Oddly, little discussion occurred over whether the other organization would actually follow through with the development. People seemed more concerned about getting the recognition for improving Chaldean Town than about the actual improvements. If their primary concern was the development of the Chaldean neighborhood then they would have expressed more interest in which group could best help the area and whether the other group could be a reliable partner capable of sharing the burden of development, rather than focusing who 'represents' the community.

More to the point, Natalie was directly told by a CARE member during an interview that one of the problems for the Chaldeans is the competition between groups, with each wanting to do things in order to get credit from the community. He had recently been approached by an organization that wanted him to help them start a program in which they would go into schools and work with Chaldean students. He told them that CARE already does this and suggested that the organizations work together, but they rejected the idea and said that they wanted to do it on their own. To him, it was clear that the other organization preferred to work independently because they wanted to be able to claim full credit for the project. Even within the Chaldean churches there is competition for control and recognition, with groups within a church competing, as well as competition between churches.

We see the competition between the organizations as a consequence of people competing to get the most gains to their reputation. The more people or groups that work on a project, the more people there are to share the credit and, perhaps, the more diluted the recognition to any one person or organization becomes. It may be that this competition for reputational benefits leads to more being done for the community than if people didn't care about recognition through an arms race of volunteer work. However, this seems to be an inefficient way to achieve cooperation because even more could be accomplished if there was cooperation between groups rather than only within groups.

A desire to be recognized as pro-Chaldean clearly motivates some Chaldeans to help their community through charity or participation in volunteer organizations. Getting this recognition has real benefits to individuals because it contributes to bolstering their reputations, which as noted leads to social and economic rewards. More cooperation takes place in the community than if this vehicle for reputation enhancement was absent. However, despite the boost to community involvement, the desire to improve one's reputation creates inefficient cooperation. Some members of volunteer organizations do as little work as is needed to get recognized as a group member and enjoy the benefits to their reputation from this affiliation. Because of the nature of the work, free-riders are difficult to detect by outsiders who only see the final product of good work being done in the community without knowing which members made it happen. Furthermore, groups compete for recognition that leads to rivalries between organizations, which may improve cooperation within a group but creates less overall success than if the groups worked together. As well, people who are primarily concerned with being recognized for their cooperation are more likely to limit their efforts to highly visible projects, leaving many important, but less public, causes unaided. Overall, cooperation in the Chaldean community is higher as a result of people getting involved for the recognition than if this motivator was absent, but this impetus also fosters free-riding and competition that hinders cooperation.

It is important to note that not all Chaldeans who engage in community-oriented work do so only for the acknowledgement and there are people who care both about their reputation and helping the community. These people are dedicated, effective and committed to their causes, and they make a significant difference to their ethnic group, and beyond. Nonetheless, a concern with reputation may increase the level of cooperation in the community by drawing in people who otherwise would not have become involved. Cooperating because of personal benefits does not make a person's contributions any less valuable.

Finally, linking back to our theory in Chapter 3 where we discussed theoretical work showing that individuals should be interested in the status of their own reputation or the reputation of the person they are helping, Chaldeans usually care deeply about the reputations of those they interact with (the HELPEE strategy). However, under some circumstances they act generously to improve their own reputation, independent of the helpees' reputations. These ethnographic findings are consistent with the experimental work on indirect reciprocity, described above.

Summary and Onward

The operation of reputation and reciprocity can be seen throughout Chaldean social life. As expected from theory, direct reciprocity emerges in dyads and long time horizons of future interactions. It seems to appear in two forms, one involving "distant friends" who engage in strict in-kind reciprocal exchanges, perhaps with the adaptive coexistence of TFT-like and GENEROUS TFT-like strategies. Prototypical examples emerge from Chaldean weddings and funerals. Culture plays a key role by (a) prescribing rules and practices that permit strict accounting, (b) supplying a default for first-time interactions, and (c) structuring interactions that might otherwise be public goods dilemmas (which are very difficult to solve with reciprocity) into a series of dyadic interactions. Cultural learning may also be contributing by helping people refine their reciprocity strategies by (for example) learning from successful Chaldeans. The second form of direct reciprocity emerges among close friends (and relatives), where cooperation still occurs in dyads or very small groups but where exchanges are frequent and use diverse currencies (e.g., baby sitting, transportation, loans). Keeping a small number of such partners in very long-term relationships allows individuals to deal with the noise inherent in such wide-ranging exchanges.

As expected from the densities of Chaldean social networks, indirect reciprocity and reputation shape much of the way many Chaldeans think about their social life. People explain how having a bad reputation has negative consequences for obtaining loans, doing business, and attracting clients, and how a good reputation can be enhanced by certain kinds of generous acts, especially those seen as contributing to the Chaldean community. The cultural linkage between generous actions in some contexts (especially for the community) and reputation appears to drive much of the behavior both within Chaldean organizations and the competition between them. Reputation also seems linked to local norms (not especially related to cooperation) about marriage, social behavior, religion, and childrearing, suggesting that indirect reciprocity, which uses "not helping" as a punishment in dyadic interactions (e.g., business loans), may be maintaining other kinds of social norms (e.g., in-group marriage).

Essential to the effectiveness of reputation in shaping Chaldean social life is the culturally evolved nature of Chaldeans' beliefs about reputation. First, many Chaldeans tend to understand a range of norm violations as generally informative about a person's character, extending violations in one isolated domain to many other aspects of behavior. For example, if a Chaldean man marries a non-Chaldean, he may be deemed untrustworthy as a business partner.

Conversely, if the same man makes a big donation to the Chaldean church, he is more likely to be deemed a trustworthy business partner. As in many societies (including, to a lesser degree, contemporary Euro-American society), this connection between behavioral domains is often an empirical mistake—behavior in one domain is not a good predictor of behavior in another domain (Hartshorne and May 1928a, 1928b; Hartshorne et al. 1930; Ross and Nisbett 1991). Of course, this statistical fact is hard to figure out in a complex world, so culture can take advantage of this ambiguity in ways that benefit the community.

The next three chapters complement and buttress these findings by further exploring the processes of cultural group selection and the emergence of social norms, punishment, and prosociality. Cultural group selection, rooted in various forms of competition among social groups that favor group-beneficial norms, can explain how cultural evolution managed to—for example—link reputation dyadic cooperation in indirect reciprocity interactions to (1) costly acts of generosity to the group, (2) in-group favoritism, and (3) in-group marriage practices. The presence of these group-beneficial practices is of little surprise given that the Chaldeans—as a social and ethnic group—survived as a Christian minority in Islamic Iraq for centuries before facing the economic challenges of Detroit (where, ironically, they are once again immersed among Islamic Arabs). It is exactly these types of practices that may have contributed to the group's survival by strengthening group identity and cooperation and by helping them to maintain their cultural integrity.

7

Social Norms and Prosociality

If someone does something bad, the whole family is shamed and has their reputation damaged.

—53-year-old male speech pathologist

When I see people going to the bars or staying out to 12 or 1 A.M., I look at them with disgust and make a dirty face at them.

—An older woman's reaction to young Chaldeans breaking
from tradition

In chapter 2, we discussed how people acquire their ideas, beliefs, preferences, motivations, values, and practices—including those related to social behavior, altruism, and punishment—from observation and interaction with other members of their social group, often using adaptive cultural-learning mechanisms such as prestige-biased and conformist transmission.[1] Building on this, chapter 3 explained (1) how cultural learning can lead to stable norms, (2) how different norms generate cultural group selection, and (3) how cultural group selection can influence genetic evolution, yielding a culture-gene coevolutionary process. Taking the first, cultural learning and social interaction give rise to social norms. If people use cultural learning (and they do), groups of individuals can culturally evolve to either highly cooperative states or noncooperative states. However, while combinations of culturally learned motivations related to altruism and punishment can lead to stable cooperation, in contexts where kinship, reciprocity, and reputation fail (even in one-shot *n*-person situations), costly punishment can still stabilize any behavior, including costly behaviors not associated with cooperation. Such combinations of stable behaviors nicely fit the phenomena commonly associated with social norms. Norms are shared patterns of behavior, beliefs, and practices that are acquired via social learning. They typically include the *behavior* (e.g., Jews do not eat pork), the *punishment* (Jews should not eat pork, and those that do should be thought less of and treated less favorably), and sometimes the *punishment of nonpunishers* (Jews should be bothered by other Jews who aren't bothered by pork-eating Jews).

133

Building from (1) to (2), we argue that cultural group selection favors the proliferation of norms that most benefit social groups in competition with other groups. Though social groups can evolve culturally to possess different social norms—which might include norms that are neutral, costly and maladaptive, or beneficial to the group—competition among social groups with different social norms will tend to spread group-beneficial norms, especially those favoring greater degrees of cooperation against other groups. This process includes economic competition, warfare, demographic expansion rates, and prestige-biased group selection.[2] (See chapter 3 for a more detailed description and a historical example of cultural group selection.)

Finally, moving from (2) to (3), although these processes have no doubt influenced recent human history (the last 10,000 years), culture-gene coevolutionary models consider how cultural group selection may have influenced genetic evolution and human prosociality over hundreds of thousands of years. By spreading group-beneficial cooperative norms involving the punishment of non-cooperative norm violators, cultural group selection may have altered the selective environment faced by genes. This altered environment may have favored genes that promote things like a readiness to acquire cooperative and punishing norms, a default bias toward helping (to avoid being punished), a preparedness to respond to punishment, and numerous other social faculties. In the rest of this chapter, we will further highlight the important theoretical patterns derived from chapter 3, then summarize some of the key experimental work, and finally, view Chaldean social life through these theoretical lenses.

This coevolutionary approach yields a number of predictions, some of which can be explored with experimental tools and others that can be seen only in ethnographic data drawn from real life. As detailed earlier, cooperation in these models is typically made possible through the presence of individuals who are willing to punish others for norm violations. The evolutionary-psychological implication of these models of the evolution of social norms is that individuals (at least some) should possess a "taste" or willingness to inflict costly punishment on norm violators. When the relevant norms prescribe cooperation or fairness, such individuals will express a willingness to punish unfairness or noncooperation. More specifically, *social norms models* predict:

a. In smaller groups (e.g., dyads) and in situations that cue local culturally evolved norms, individuals will possess a non-trivial, positive baseline preference for punishing norm violations, even in one-shot interactions. By "positive baseline preference" we are referring to those motivations to punish that are not accounted for by models of kinship, reciprocity, or reputation. However, as explained in the previous chapter, opportunities for reputation building and repeated interaction can increase the amount of punishment observed because punishers can, under some circumstances, increase their direct benefits by punishing unfair or noncooperative players early on in a repeated or in a reputation-related context. In these situations, evolved motives arising from social norm coevolutionary processes (the nonzero baseline), repeated interaction, and reputational effects all combine to produce the observed behavioral patterns. As noted in the last chapter, experimental results comparing one-shot Ultimatum Games that allow for reputational information with those that do not allow for it show the predicted positive baseline taste for punishment, which does not decline with several rounds of experience.

b. In larger groups (e.g., $n \geq 4$), the strength of this taste for punishment will not depend very much on the presence of repeated interactions, only on the non-zero baseline. The evolutionary forces in social norms coevolutionary models do not favor punishment because evolutionary forces do not increase cooperation with the punisher. That is, people punish *despite the fact* that punishment will not increase cooperation with the punisher. In culture-gene coevolutionary social norm models, punishment evolves because social groups with more punishers can maintain more cooperation than social groups with fewer (see chapter 3). Neither repeated interaction nor reputation are required in n-person situations, which is good because we know that neither is effective at maintaining cooperation in such circumstances. Ergo, punishment in one-shot cooperation games should not be much different from punishment in repeated n-person games.

c. Individuals will be willing to punish norm violators even when punishment is costly to the punisher in time, money, reputation, or effort. In cases of cooperative norms, people will punish according to the violator's deviations from the expected degree of cooperation or fairness. This "expected degree" is culturally learned.

d. The effectiveness of punishment depends on widely shared information about who did what to whom. Such widely shared information spreads the cost of punishing out over many individuals, and allows those most able to inflict high costs on the violators at low cost to themselves to do the punishing. Gossip provides an important medium for this dissemination.

e. Social norms may take advantage of our kin psychology by punishing (damaging the reputation of) close kin. Because one's family may be more vulnerable to punishment than the norm violator himself, punishment that threatens the violator's family can be a powerful form of norm enforcement. This threat will put an additional burden on families to keep their members in line, leading them to punish their own kin. People will punish their kin by means other than gossip since they won't want others to know about the norm violations.

f. People should be quite responsive to punishment, or the threat of punishment. That is, psychologically, we should expect that there might be "punishers" out there and react accordingly.

g. If cooperation or fairness is the local normative expectation for a particular situation, then some people will be fair and cooperative, even if no opportunity to be punished exists. That is, some people will have internalized the social norms (Gintis 2003a, 2003b; Gintis et al. 2003).

Repeated interaction, however, in which others are consistently defecting without punishment will cause these individuals to revise their beliefs about of the normative expectations in a particular context.

The above discussion of how gossip and reputational information allow focused punishment might lead people to conflate indirect-reciprocity models that use reputation to stabilize n-person cooperative interactions, such as that studied by Panchanathan and Boyd (2004), with these social norms models, which use costly punishment. The key difference is that the social norms models require that individuals engage in costly punishment acts, whereas the indirect-reciprocity models allow individuals to perform a defection on individuals with bad reputations. In real life, we think both are certainly important, although they can often be difficult to distinguish (thus the experimental data are critical).

In the next section, we will examine the experimental support for these predictions and briefly consider some alternative evolutionary interpretations. This is followed by a section that explores the application of these predictions to Chaldean social life.

Experimental Foundations for Social Norms

Public Goods Games

Experiments called Public Goods Games capture the essence of the n-person cooperative dilemma. In a prototypical game, four anonymous players are each endowed with $10. They can each contribute (without knowledge of the others contributions) between $0 and $10 to a project. The total amount contributed to the project is doubled (or multiplied by values such as 1.5) and distributed equally among the four players. For example, if all four players contribute $10 to the project, this $40 is increased to $80, and then divided such that each player goes home with $20. Alternatively, if three people contribute the full $10 to the project and one person gives nothing, $60 is shared equally among the four players ($15 each). This means that the three who contributed $10 each go home with $15, and the one who gave zero goes home with $25. Finally, if all four players give nothing, everyone goes home with $10. If the game is one-shot for the four players (not repeated with the same people), purely self-interested individuals will always contribute nothing. Many experiments using this basic setup among university students show that *mean* contributions in one-shot games are between 40 percent and 60 percent of the players' initial endowment (Dawes 1991; Ledyard 1995), although individuals typically offer either their full endowment or nothing. Thus, in accordance with prediction g above, some people will cooperate, at least initially, in one-shot interactions. If asked why they cooperate, many people say it's because they believe it is the "right thing to do" (Henrich and Smith 2004). It should also be noted that the results of Public Good Games, as expected, vary across populations (Henrich et al. 2004).

For our purposes, two variations on the basic Public Goods setup are important: repeated interaction and punishment. What happens if players in this game play with each other repeatedly? That is, what happens if the four players make contributions, their contributions are increased, equal divisions are made, payoffs are received, and then the same people do it all over again for several rounds? Figure 7.1 shows the effect of playing the Public Goods Game repeatedly (Fehr and Gächter 2000). In this experiment, players were endowed with 20 MU each round ("monetary units" convert directly to currency). In the first round of this "Partner Treatment" (playing with the same people), players contributed an average of about 9 MUs (45 percent of their endowment) to the group. The triangles from rounds 1 to 10 show a gradual decline in the average contribution. By round 10, players are contributing an average of only 2.8 MUs (14 percent of their endowment). This, and many subsequent replications, make two important points: (1) the possibility of repeated interaction does not substantially increase contributions in round 1 (vis-à-vis the one-shot game), and (2) repeated interactions drive cooperation down as self-interested individuals exploit cooperators. Thus, as predicted by social norms approach (prediction *b*), repeated interaction in *n*-person group does not lead to stable cooperation (see Bendor and Mookherjee 1987; Boyd and Richerson 1988; Joshi 1987).

Building on the basic game just described, Fehr and Gächter (2000) added the ability for players to pay some of their endowment to take money away from other players. That is, the experimenters added diffuse punishment. Figure 7.1 shows what happens when punishment is added after cooperation has deterio-

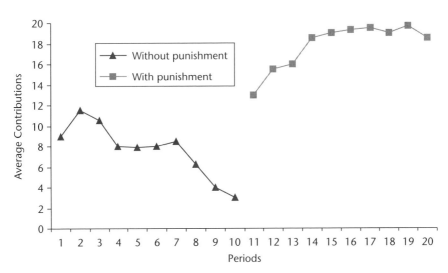

Figure 7.1 This shows the effect of playing the Public Goods Game repeatedly over ten rounds with the same people. Rounds 1 to 10 show the game without punishment opportunities. Rounds 11 to 20 show the game with punishment possibilities. Here, in each round, the contributions of all players are multiplied by 1.6 and distributed equally among the four players. Adapted from Fehr and Gächter 2000.

rated during rounds 1 to 10. Initially, in round 11 (the first punishment round), contributions spike up to 13 MU (from 14 percent to 65 percent) before anyone has actually done any punishing. Note that this is above where the same players started in round 1. Such an effect is consistent with the idea that people anticipate the presence of punishers and adjust (prediction *d*). Second, mean contributions are driven up, by punishment and the threat of punishment, to nearly full cooperation—almost everyone gives all of his endowment (over 90 percent).[3] This general prediction, for punishment to stabilize cooperation in *n*-person situations, has arisen directly from theoretical work on the evolution of social norms (in particular, see Boyd et al. 2003; Boyd and Richerson 1992a; Henrich and Boyd 2001).

Are the individual-level patterns of punishment consistent with our theory? Do players in this game punish individuals who deviate substantially from the cooperative norm more than those who deviate only a little (prediction *c*)? The dark bars in Figure 7.2a show the average amount of punishment administered by players according to the deviant's deviation from the average contribution of their group. For example, if a player contributed between 14 and 20 MU less than the group average, punishers paid an average of almost 7 MUs to punish such deviants. Clearly, people preferentially punished greater norm deviants, confirming prediction *c*.

A skeptic, perhaps one wishing to defend the acultural origins of human nature, might suggest two other strategic reasons why a player might punish another in these games. First, in a repeated game, an individual might punish in the early rounds in an effort to get others to cooperate so he can defect on them in later rounds (when he would neither cooperate nor punish). Figures 7.2a and b address this possibility. In Figure 7.2a, the grey bars show the willingness of individuals to punish others in a Stranger Treatment. The Stranger Treatment is very similar to the Partner Treatment (where individuals play with the same people for the entire game) except that now individuals are randomly rematched every round so that it is very unlikely that they will encounter the same individuals again. This change removes the incentive to punish in the earlier rounds. As illustrated, the pattern of punishment is virtually identical, confirming that punishments are not driven by opportunities for repeated interaction (consistent with prediction *c*). Second, using data from a very similar experiment in which players were guaranteed never to meet the same other anonymous players during a six-round game (people played six one-shot games), figure 7.2b shows the same pattern of punishment and demonstrates that punishment does not decline in the last few rounds. Moreover, figure 7.3 shows that in this situation, punishment can still lead to substantially more cooperation, despite the nonrepeated nature of the games. That is, people punished even when the game was really a series of one-shot interactions.[4]

A skeptic might propose a second explanation: by punishing noncooperators, selfish individuals may be able to reduce their relative fitness deficit (Price et al. 2002). This idea is both theoretically flawed and empirically unsupported. Theoretically, careful mathematical analysis using tools from game theory show that this logic will not support the evolution of punishment unless human

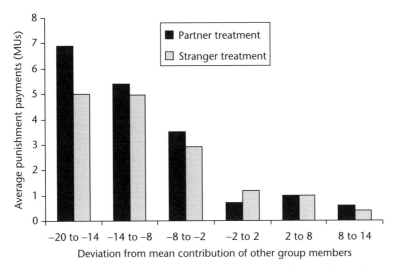

Figure 7.2a For the repeated game involving partners, the dark bars show the average amount of punishment administered by punishers to noncooperators according to their deviation from the group mean. For the "stranger treatment," the grey bars show the average punishment payment. Adapted from Fehr and Gächter 2000.

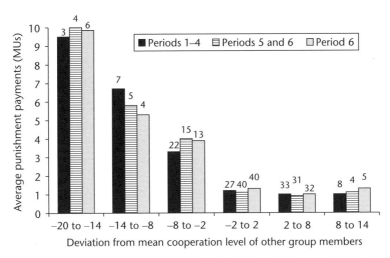

Figure 7.2b Shows data for six one-shot Public Goods Games. This figure plots the average punishment payment (vertical axis) delivered to players according to their deviation from their group's mean contribution to the public good in that round. This is broken out by rounds: rounds 1–4, 5–6, and 6 alone. Adapted from Fehr and Gächter 2002. The numbers above the bars are the percentage of all cases represented by the bar. For example, 3.1 percent of all cases deviated between 14 and 20 MU below the average cooperation of other group members in periods 1–4.

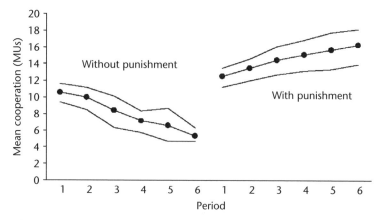

Figure 7.3 The difference between a series of *one-shot* Public Goods Games, with and without punishment. In the experiments, individual contributions to the public good were multiplied by 1.6 and distributed equally among the four players in the group. Punishers could pay to punish at a cost of 1 MU to subtract 3 MUs from the punishee. Adapted from Fehr and Gächter 2002.

groups are very small (Gintis et al. 2003; Huck and Oeschssler 1999). The reason is simple. By punishing noncooperators to reduce their fitness advantage, punishers create a fitness deficit between themselves and individuals who cooperate but don't punish (and thereby don't suffer the costs of punishing). Empirically, Fehr and Gächter (2005) have replicated the Public Goods Games studies above in situations where players can pay 1 MU to punish 1 MU. In such circumstances, punishing cannot alter relative payoffs; thus the above explanation predicts zero punishment. In contrast, the empirical results decisively show exactly the same patterns of punishment and cooperation seen above, although somewhat reduced.

Moreover, the evidence from Public Goods Games indicates that, as expected from social norms theories, people have both a taste for punishment and an expectation that punishers are out there. Contrary to what alternative evolutionary hypotheses predict, this taste is for punishing norm violators (not merely noncooperators) and applies even to one-shot situations in which reputation-building is impossible. Coevolving with the "taste for punishment" is an expectation of being punished for norm violations (like going naked to a wedding). No other class of evolutionary theories makes these predictions.

In addition to Public Goods Games, there is another important set of recent experiments that test the predictions made by social norms theories. Third-Party Punishment experiments directly measure the willingness of individuals to punish others for unfairness (norm violations) in interactions that the punisher was not involved in. In the simplest version of the experiment, there are three players, A, B, and C. Players A and B are allotted a sum of money (say 100 MUs), and it is player A's job to divide the money between players A and B. A can give B whatever amount A wants to, from nothing to the full endowment.

Player C is endowed with 50 MU, and he can pay any amount of his endowment (from 0 to 50) to take money away from player A. For every 1 MU that C pays, 3 MUs are taken from A, and this money disappears (it does not go to player B), so no one gains financially if C punishes A. As usual, the money is real, the players are anonymous, and the games are one-shot (no repeated play). Figure 7.4 illustrates the punishing results from this experiment: the diamonds show the relationship between what A gave to B, and how much C was willing to punish A. For example, the plot shows that if player A gave 0 MUs to B and kept 100 for himself, player C would pay an average of about 14.5 MUs to take 43.5 MUs away from A. As expected, the amount of punishment declined as the deviation from the normative expectation of a 50 MU offer decreased. Third parties—who are not involved in the interaction—will pay to inflict punishment on others who violate normative expectations (giving half in this case). Moreover, the squares on figure 7.4 show that not only will people punish but that others have well-formed beliefs about the willingness of third parties to inflict costly punishment.

Finally, because social norms theories arise from the dual inheritance of culture and genes, the norms and expectations to which these aspects of our psychology get attached will depend on local cultural evolution. This means that tastes for punishment, equity, and cooperation—as measured in experiments like the Ultimatum, Third-Party Punishment, and Public Goods Games—should vary culturally, depending on and reflected by local norms. All three of the games mentioned have been extensively tested in small-scale societies, including places as diverse as Mongolia, the Amazon, Tanzania, and Papua New Guinea. This work in small-scale societies confirms that although the broad patterns are consistent, substantial differences in the magnitude of punishment and the willingness to cooperate or offer equitable distributions

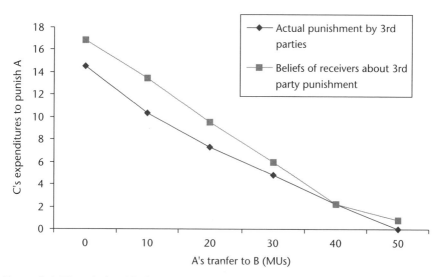

Figure 7.4 The relationship between what Player A gives to Player B and the amount that Player C is willing to punish. Adapted from Fehr and Fischbacher 2004.

varies greatly. Cooperative contributions in the Public Goods Games, for example, range from a mean of 22 percent among the Machiguenga horticulturalists of the Peruvian Amazon to a high of 65 percent among the Aché foragers of Paraguay. Ultimatum Game offers range from about a quarter of the total among Quichua and Machiguenga in the Amazon to 58 percent among the whale hunters of Lamalera (Henrich et al. 2004). Similarly, people's willingness to punish varies from near zero to the levels we observed among adults in industrialized societies. Very recent work in small-scale societies has replicated, reinforced, and extended these original findings, whereas work in Russia using Public Goods Games confirms that the ability of punishment to stabilize high levels of cooperation is culturally dependent—punishment can stabilize intermediate levels of cooperation, too (Gächter et al. 2005).

Wason Selection Task Supports Coevolutionary Theory

Other experimental evidence for the social norms approach, which has often been misinterpreted as supporting reciprocity-based theories, comes from experiments using the Wason Selection Task to study the effects of content and context on reasoning (Cosmides 1989; Gigerenzer and Hug 1992; Stone et al. 2002; Sugiyama, Tooby, and Cosmides 2002). In this task, subjects have to solve a logic problem (of the "if p then q" sort). In the first condition, the problem is framed either without context or with a context that does not involve someone cheating on a social norm. In the alternative condition, the subject is given a problem with the identical abstract logical character as the first, but that is now framed as the violation of a social norm. In this condition, solving the logic problem is equivalent to identifying whether someone is violating a social norm. The robust result is that although the logical character of the two kinds of problems is identical, subjects find the social norm problems substantially easier to solve. For example, in one experiment, only 25 percent of subjects answered correctly in the first condition, whereas 80 percent got it correct in the second. Interestingly, this striking difference is not affected by whether the responder is framed as the person being cheated or as a third-party observer. Cosmides and colleagues explain this result by positing that humans have an innate decision-making module devoted to detecting cheaters on social contracts. We agree! Social norms coevolutionary theories predict that humans ought to possess the psychological skills, motivations, and preferences for learning social norms, identifying norm violators ("social contract violators," to use Cosmides's lingo), and punishing those who substantially deviate.

However, we maintain that social norms theories provide a better, more complete, explanation for these results than do the reciprocity-based theories that have served as the theoretical foundations for the interpretations of Cosmides and her coauthors. First, reciprocity-based theories do not deal with the existence of culturally learned social norms at all. Where do these "social contracts" come from in the first place, and why are people readily able to socially learn new ones, even unfamiliar ones? Thus, it remains unclear how Cosmides goes from theories of reciprocal altruism to "social contracts." Nothing in reciprocity-based theories leads to arbitrary social contracts without cultural transmission. To emphasize the point: the experimental materials are not generally the kind of cooperative dilemmas associated with the evolution of cooperation via reciprocity; the situations involve all kinds of seemingly arbitrary social norms, ranging from enforcing a drinking age to tattooing one's face before eating a certain food. Theories of reciprocity are silent on such noncooperative situations. The social norms model, however, provides the only evolutionary account of the emergence of social norms and predicts that individuals should be psychologically "geared up" to readily acquire seemingly arbitrary noncooperative norms (because, as we have emphasized, culturally learned punishment can stabilize both noncooperative and cooperative norms).

The second reason why social norms theories are better able to explain the Wason Selection Task results is slightly more complicated. Consider two facts from reciprocity theory. First, reciprocity-based theories do not predict that people ought to punish third-party norm violations, even though we observe this ethnographically and experimentally. In fact, there is no reason to expect people to be able to detect norm violators. Unfortunately, some have claimed that reciprocity-based theories predict that people ought to monitor social contracts for violations in order to learn about potential future partners. But under reciprocity, seeing someone defect (presumably this is akin to breaking a social norm in the minds of those who make this argument) does not tell you whether the person is using a TFT-like strategy because TFT-like strategies are contingent cooperators only. Such strategies should cooperate only if the expected number of future interactions (see chap. 3) is sufficiently high, the cost/benefit ratio (c/b) is sufficiently low, and the number of players (N) is also low. As shown in figure 6.1 for a two-person interaction, "violations of social contracts" should be observed and expected all the time in a world governed by direct reciprocity. Players using TFT should expect other TFT strategists to defect anytime a combination k and c/b places one outside the gray region. In essence, there are no social contracts or norms in a world of direct reciprocity, only c/b ratios and the shadow of the future (k).

Second, in the large anonymous situations tested by Cosmides and others, reciprocity theories would (if properly applied) predict that people's cheater-detection equipment should be "switched off," because reciprocity could never stabilize cooperation in such large groups. Therefore, Cosmides's findings actually violate a clear prediction about direct reciprocity derived from years of rigorous mathematical analysis.[5] In contrast, social norms theories are specifically constructed to incorporate culturally transmitted preferences or strategies for third-party punishment and, as explained above, predict the emergence of social norms, third-party punishment, and the associated psychological machinery.

Theoretical Patterns in Ethnography

The above experimental data all fit social norms theories quite well. However, it is very important to ask whether all this is just "games in a lab" or whether these theoretical, and now experimental, patterns play out in real life. Below, we bring

Chaldean ethnographic data to bear on several issues. First, social norms theories predict that punishment will maintain all kinds of norms, cooperative and otherwise. Consequently, we will first explore how normative punishment operates in Chaldean life. Are there noncooperative social norms that are maintained by third-party punishment? This is important because noncultural theories cannot explain such punishment. Second, because the psychology of social norms not only coevolved with culture but actually *required* culture in order to evolve, we should ask: Does punishment in cooperative or public good situations depend on the culturally transmitted salience of the public good, and less on the details of the costs and benefits?[6] If culture influences social behavior, then noncultural theories are deeply insufficient.

Punishment Can Stabilize Any Norm, Not Just Cooperative Norms

One of the values of studying formal mathematical models of evolution, social behavior, and culture is that they frequently produce nonintuitive or unexpected predictions that extend beyond the properties initially under investigation. One of these unexpected extensions occurred when researchers included punishment in cultural evolutionary models of cooperation. They found that the benefit, b, delivered by the cooperator drops out of the final equation, leaving only the cost, c, of doing the cooperative behavior and the costs associated with punishing and being punished. The conclusion from these studies is that the right balance of punishment can lead to the maintenance of costly behavior in a group regardless of whether that costly behavior delivers benefits to anyone. In studying cooperation, the authors had arrived at a more general explanation for norms that could explain cooperation (when $b > c$), and a number of other puzzling—but empirically rampant—behavioral patterns (which we have called norms). In fact, even small amounts of conformist learning can generate enough punishment to stabilize some fairly costly norms (at least in the short term); these are norms that cost both the individual and the group (Boyd and Richerson 1992; Henrich and Boyd 2001).

Among the Chaldeans, given that punishment is integrated with and targeted mostly through gossip, ethnographic experience suggests that a fear of getting a bad reputation leads not only to compliance with cooperative norms but to compliance with any norms that are widely accepted by the community. To illustrate the power of punishment and its role in conformity, we consider two situations in which Chaldeans get punished for breaking norms: marriage and caring for elderly parents.

Norm of Endogamous Marriage: Chaldeans Should Marry Other Chaldeans

Some of the most vehement reactions and punishments apply to people who have married outside of the ethnic group, although the extent of the reprimand depends on the type of non-Chaldean the person marries. Though people may gossip about marriage to a non-Chaldean "Catholic," a person is more likely to be disowned and made an outcast if he or she marries a "Jew," "Muslim," or "Black." Everyone agrees that there will be some gossip anytime a Chaldean marries a non-Chaldean, though a fair number of people believe that the amount of gossip is declining as the frequency of exogamous marriages increases (though it remains small). In some families, the punishment is much worse for the first child who marries outside of the ethnic group, with relatives accepting subsequent marriages to non-Chaldeans by other children more easily. Sue said that her family accepted her marriage to a non-Chaldean quite well, although people in the community regularly ask her, "How long have you been married?" because they cannot believe that her marriage will last. She explained that her family had already gone through her sister marrying a Jewish man, and that this eased the way for her. Her sister was not so lucky. She was disowned by her mother, who wouldn't go to the wedding and said that she didn't want any of the other children to attend the wedding. The mother didn't speak to the sister for two years, but everyone gets along well now, and the husband has been accepted into the family.

Khalid, a young man who had dated a Muslim woman, said that his family warned that if he married her, "I'd be on my own." He also said that in most families, if someone marries a "Black," "Muslim," or "Jew," he is "kicked out of the family" and the family feels "disgraced." In less extreme cases, such as marriage to a non-Chaldean "Catholic," most of the reactions take the form of gossip. However, Khalid said that even when a person marries a non-Chaldean who is Catholic, the parents will often not attend the wedding, and, consequently, other relatives will also not attend. Members of the community ask the person who married a non-Chaldean, "Why didn't you marry a Chaldean?" and among themselves, they ask, "Why didn't he find someone from our own culture?" or say, "What a waste. Why did he go to a stranger?" What is the cost of being talked about, or having people question why you chose your spouse or wonder how long your marriage will last? Such measurements are nearly difficult to make, but clearly, marrying a non-Chaldean does hurt a person's reputation and that of his family. The effect of this on decision making likely helps explain the low rate of exogamous marriage among Chaldeans (chap. 4). A fear of punishment may cause members of the community to conform to the group's expectations and beliefs about what constitutes a good marriage partner.

Damaging reputation through gossip can be an incredibly efficient system for channeling punishment through those most able to deliver it. In the case of marriage norms, community members gossip about a person (who is marrying a non-Chaldean) and his family. This gossip damages the reputation of the family, which can lead to all kinds of direct losses (jobs, loss of special "Chaldean credit" for store merchandise, valued marriage partners for other children, etc.) down the road. Such third-party punishment, or the threat of it, compels the family to rebuke the child for the norm violation. The child, who may be convinced that his best option lies with marrying a non-Chaldean, will endure

(1) damage to his own reputation, (2) damage to his close kin (this plays on kin psychology), and (3) direct punishment from his own kin.

Ethnographic description of this sort does not allow us to argue that this "norm of endogamy" harms the individual and benefits the group, or harms the individual and harms the group. Nonetheless, it seems difficult to argue that it always benefits the individual who wants to get married outside the group. There is little doubt that some individuals can benefit by expanding their pool of eligible partners by a factor of 50 (100,000 Chaldeans to 5 million Detroit residents). Thus, with a marriage rule that is universally applied to all Chaldeans, the marriage rules cannot be generally benefiting all individuals, and marrying endogamously is not readily interpreted as a cooperative dilemma. However, groups that enforce endogamy may persist because those groups who possess such norms remain in existence longer than groups lacking them. Imagine, for example, a group that had a general norm (applied to males and females) of group exogamy. The social group would rapidly disappear as a culturally identifiable unit (without some other additional mechanisms to maintain coherence). Furthermore, as we'll discuss in the next chapter, a norm of endogamy may influence how people think about their group in relation to other groups, which may benefit the group but does not necessarily benefit the individual.

Norm of Care of Elderly Parents

The threat of punishment and potential damage to one's reputation may also contribute to children caring for their elderly parents. Assuming that the children are able to help their parents, Chaldeans consider it truly terrible for children to neglect this responsibility by, say, failing to look after their parents in the parents' home or failing to bring their parents to live with them. Unless the parents require hospitalization, putting them in a nursing home does not constitute caring for them. Having parents in a nursing home is a source of shame.

When a child does not care for his parents, the extent to which others perceive him as bad varies. In the eyes of some, this offense destroys a person's reputation, whereas for others, it leads to disrespect, but not to a cessation of all interactions or to a dismissal of the person as "evil." Most Chaldeans say that this offense is rare, but if it happened, people would gossip about the children, consider them to be selfish and disrespectful, and generally think poorly of them. Most agree that if they were close to the children, they would make comments and criticisms directly to them, such as "Shame on you, how could you do this?" or "How dare you not take care of your parents!" But more distant friends and acquaintances would only gossip behind the children's backs. Among the more extreme reactions, a couple of men said that if the child was someone they didn't know, then they would go out of their way to avoid meeting him, and one man said that he would treat the person "like a dog." Another man said that the child would become "an outcast . . . people would think there's something sickening about him." Consistent with the findings of Fehr and colleagues above, people's taste for punishment varies, and strong tastes for punishment link directly to powerful negative emotions, such as anger and disgust.

The potential damage to a person's reputation, and the possible costs such as a loss of respect, exclusion from the community, and termination of friendships could be a large motivation for children to take care of their aged parents in a particular way and may lead to the perpetuation of the Chaldean tradition of children taking on the direct responsibility of parental care. We are not suggesting that children only care for their parents out of a fear of punishment. In such close-knit families as are common among the Chaldeans, strong feelings of love and respect surely contribute to the care giving. However, non-Chaldeans also love their parents and yet, many of the same social class in the same area do not bring their aged parents into their home to care for them, nor do they feel that it is villainous to put them in a nursing home.

This is a norm about the "right way" to bear the cost of aging parents. It's not about group cooperation or group benefits, although it is individually costly. Why would someone (a third party) care how a non-relative treats his parents? The social norms approach is the only evolutionary theory that proffers an explanation.

Fear of Gossip and Conformity

Gossip and shunning can be so pervasive that people feel they have no choice but to conform to the norms of the group because the punishments for doing otherwise are too great. We mentioned earlier that Jennifer said she could never date a black man because of the repercussions to her and her family. She went on to say: "Our world revolves on gossip. Everything you do is affected by what people will say. Who you date, what you wear to church, what you say, is all influenced by what others will say about you. Apart from not liking to be talked about, there are other repercussions. People will look down on you and your whole family."

Before she does or says anything she thinks about the consequences. "It's always in the back of your mind that people will talk about you." Jennifer described the necessity of acting in a way that the community perceives as appropriate, rather than in a way that feels comfortable. A Chaldean "can't be halfway into the community," she said. "You can either be completely in the community, in which case you're part of it, and you do what is acceptable; or you completely distance yourself from the community. It's not worth it to go against the group and what's acceptable. Just go with the flow. It's too hard to be different."

Another woman, who moved to the United States when she was 16, also spoke of having to modify her behavior to fit into the community. She said she is more at ease with "Americans" than Chaldeans. "With Americans I can be myself, say anything, tell a dirty joke," she said. "With Chaldeans, you have to be very careful; they critique everything. You have to watch what you say." She said she feels that she has to be different from who she really is in order to get the approval of the community.

Our interview data suggest that people modify a wide range of their behavior to conform to the local Chaldean norms. Much of what is mentioned has little to do with cooperation, yet a fear of third-party punishment through gossip is an ever-present specter that looms over individuals who want to be part of the group. We should note that, in most other circumstances in human history, people cannot easily leave a group; it's only the fluid anonymity of contemporary American society that permits one the option of distancing oneself from the community.

It is important to realize that indirect reciprocity does not provide an explanation for the effect of gossip and reputation on these kinds of behaviors. Indirect reciprocity demands that behavior be cooperative: the benefit, b, received in an interaction must be greater than c, the cost of delivering it. But the normatively prescribed behaviors mentioned above provide no benefit, and they have costs only for the doer. Indirect reciprocity will not stabilize nonhelping behaviors. Social norms, however, can address both the presence of such noncooperative norms and, as we'll discuss, the need for gossip to remain at least somewhat reliable (indirect reciprocity requires high-fidelity gossip, but individuals have an incentive to spread false gossip).[7]

Third-party punishment is most effective when it is diffuse and applied by those best able to inflict the most damage at the least cost. This demands a mechanism for spreading information about norm violations, and getting that information into the hands of the punishers best able to efficiently inflict costs. Among the Chaldeans, one of the functions of gossip in the community is to identify norm violators and damage their reputations. People spread details about the violator's transgression, potentially damaging the person's reputation. Others respond to the gossip and punish the norm violator.[8]

Punishing in Public Goods Situations Depends on Culturally Transmitted Salience

To probe how inclined Chaldeans are to engage in third-party punishment, Natalie described to a subset of people two hypothetical public goods situations involving a potential norm violation that wouldn't directly hurt them and asked each person (1) what he thought of the person in the scenario, (2) whether he would say anything directly to the person about his action, and (3) whether he would tell anyone else about the person's action. These scenarios were also designed to gain information on behaviors in different cooperative domains, to determine whether punishment varied by domain, and to see if people would likely confront someone when punishing her or would punish her via gossip and damage to her reputation. Here, both direct confrontation and gossip are third-party punishments because the punisher was not directly affected by the acts.

The hypothetical situations were the following:

1. A coworker finishes drinking a can of soda. There is a garbage can next to him and a recycling container in the next room. The coworker throws the empty can into the garbage rather than the recycling bin.[9]

2. You're at mass, and a collection plate is going around to raise money for Chaldeans in Iraq. You are sitting next to a man who you know is very wealthy. You notice that when the collection comes to him, he doesn't put in any money.

In the first scenario, 67 percent of the people ($n = 17$) said that they would punish the non-recycler by directly scolding him, usually in the form of questioning why he wasn't recycling and pointing out that recycling is available. Some people explained that this situation has actually occurred and that they have said things such as "Why do you put it in the garbage when you have another place to put it?" or "Hey man, what are you doing? Why don't you recycle that?" or simply "There's a recycle bin in the other room." However, only 22 percent of respondents said that they would form a negative opinion of the person, and only 11 percent would tell anyone else about the person's failure to recycle (fig. 7.5). In this case, most people would engage in third-party punishment, but the punishment would be directed right at the offender, and there would be minimal consequences in terms of damage to reputation. People do seem to feel the need to step in and scold the non-recycler but the cost of the punishment seems slight. If a responder had a negative opinion of the person, it was usually that the non-recycler was somewhat lazy, not that he was a bad

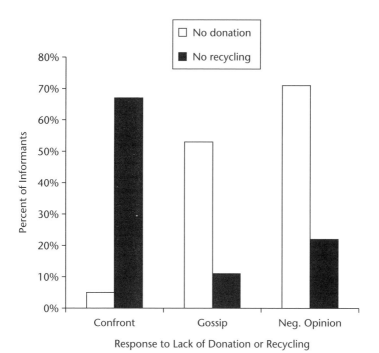

Figure 7.5 Reaction to (a) a coworker who doesn't recycle ($n = 17$), and (b) a wealthy man who doesn't make a donation to Chaldeans in Iraq ($n = 20$).

person. This didn't seem to be a behavior that people got very heated about, with several people saying they don't care about recycling at all. As a social policy issue, recycling is often considered a quintessential public goods problem.

In contrast, reactions were markedly different for the scenario involving the wealthy man who doesn't contribute to the collection for Chaldeans in Iraq. In this situation, only 5 percent of the people would say anything to the non-contributor, but 53 percent (10 times more) said that they would tell other members of the community about the man's failure to donate. Further, 71 percent (3+ times higher than in the recycling example) would form a negative opinion of the man, with most people thinking he is cheap, one person saying that he is a "bad man," and another saying he would generally think poorly of the man (fig. 7.5). Of the people who said that they would not form a negative opinion, nearly all said that they couldn't condemn him because he may have donated elsewhere, suggesting that if they knew he hadn't contributed anywhere, then they would think negatively of him. Interestingly, it is also in the donation scenario that most people would develop a negative opinion of the man, whereas in the recycling case, less than a quarter of the people would think ill of the non-recycler.

Based on these two examples, it seems plausible that when the offense is serious and people feel that a real wrong has been committed they rely more on gossip than on direct confrontation. This serves two purposes. First, it lets more people know about the offense so that the subsequent punishment can be greater. Second, it may be less costly to gossip than to confront the person when the issue is serious because the relationship between the people can be damaged if the scolding gets heated and the offender reacts badly to the criticism (punishment is costly). In the case of a minor infraction, such as recycling, the punisher can make a simple informational statement such as "There's a recycle bin in the other room." That lets the person know that you don't approve of his failure to recycle, but no real confrontation develops and, although the non-recycler may be embarrassed, he is unlikely to feel deeply injured by the comment. However, when people are dealing with such a culturally salient and moralized issue as charity for Iraqi Chaldeans, the risks involved in direct confrontation increase, and gossip is the weapon of choice.

Thus, it appears that there are at least two strategies for third-party punishment. For minor infractions, punishment is aimed directly at the offender because (1) the cost is relatively low, and (2) gossip is unlikely to be effective because a minor infraction lacks the generally recognized cultural significance to spread very widely or cause gossip receivers to take any action (to punish). In the recycling case studied above, gossip about non-recycling would likely neither spread nor damage any reputations. However, for serious infractions, the third-party punisher relies on gossip, which creates potentially high costs for the offender because the cultural salience of the offense prompts people to both spread the gossip and form negative opinions (which informs future behavior) once they hear the gossip.

Do Chaldeans Alter Their Behavior Because of the Threat of Third-Party Punishment? Do They Internalize Some Norms?

To what extent does the threat of third-party punishment motivate people to conform to group norms? To assess this, Natalie posed two more hypothetical situations and asked people to predict how they would behave in each of the scenarios:

1. A friend of yours is walking down the street, and she has a food wrapper in her hand that she wants to throw out. She looks around for a garbage can, but there isn't one around. She looks around again and notices that there is no one else on the street. Do you think she'll drop the wrapper or keep it with her until she finds a garbage can?
2. The same friend is walking down the street with a food wrapper in her hand that she wants to throw out. She looks around for a garbage can but can't find one. There are people out on the street with her, and she notices that people are watching her as she walks by. Do you think she'll drop the wrapper or keep it with her until she finds a garbage can?

In earlier questions about littering (and in contrast to recycling), Chaldeans responded passionately; the impression is that Chaldeans are very committed to the public good of keeping common areas clean. Because there is evidently a shared belief that littering is wrong, we suspected that no one would admit to littering, which is why in scenarios A and B Natalie asked people what they thought someone else might do in these situations. Fifty-five percent ($n = 17$) of respondents said that the person would not litter in either scenario. These people seem to be unaffected by the possibility of punishment in this situation, being motivated primarily by the belief that it is wrong to litter and not requiring a threat of punishment to induce this cooperative behavior. However, 45 percent of respondents said that they thought a friend would or might litter if no one were around to see it. Of those, nearly 90 percent said that the friend would not litter if people were watching. This seems to indicate that for those who are not motivated to cooperate by a belief in the rightness of the norm or rule, a threat of third-party punishment will induce compliance—because they believe punishers are out there. Generalizing from this example, it appears that for people who are not motivated to conform because of internalized beliefs about the rightness of a norm, third-party punishment is an effective mechanism for inducing conformity and cooperation, and it seems plausible that costs associated with this form of punishment are sufficient to induce normative behavior. Overall, these data are consistent with the experimentally observed effect of punishment on cooperation in repeated public goods games.

Heterogeneity in the Internalization of Norms

Both theoretical and experimental work suggests that individuals vary in the degree to which they (1) adhere to a norm without the threat of punishment and (2) punish norm violators (Boyd et al. 2003; Boyd and Richerson 1992; Fehr and

Schmidt 1999; Fehr and Fischbacher 2003). Some people seem willing to adhere to costly norms regardless of who is watching and are willing to punish norm violators. Others adhere to norms but won't punish violators. And still others will only adhere to norms in the presence of punishers, and they may or may not punish other violators. Thus, ethnographically, our question is whether some cooperative norms or behaviors are internalized. Is fear of punishment the only reason for adherence to cooperative norms? To investigate this, Natalie asked 17 randomly selected Chaldeans whether or not they participate in three public goods, and why. These public goods involve society as a whole, not just the Chaldean community, and violating the rule may result in informal punishment from coethnics or formal punishment from the police. These rules are: obey jaywalking laws, obey stop signs, and obey littering laws. Without taking into account what we know about Chaldean culture, we see no way to predict which of these public goods would be associated with more internalization and/or punishment. Given what we know about Chaldean culture in metro Detroit, we expect a priori that littering would have both the highest internalization and the highest compliance.

Jaywalking

Natalie asked the people whether they ever cross at a red light when there were no cars coming. Seventy-one percent said that they do cross against a red, and 29 percent said that they never break this rule. Are the 29 percent afraid of punishment, or have they internalized the rule against jaywalking? Our follow-up questions indicate that those who don't cross against a red are worried. One man said that he follows this rule because of a fear of punishment, noting that jaywalking is against the law and that he could get ticketed. The four other people who don't jaywalk said that the only reason is the risk involved—it's too dangerous. This response, while not technically "punishment," is certainly not evidence of an internalized norm. Interestingly, a woman who said that she does jaywalk claims she does so because it is not illegal here (though it is), implying that she wouldn't jaywalk if she could get in trouble for it. Overall, jaywalking has low compliance, and nothing suggests that people have internalized beliefs about it, despite the fact that it is the local law.

Obeying Stop Signs

People were asked if they have ever gone through a stop sign late at night and after they have slowed down enough to see that there are no other cars at the intersection. Twenty-nine percent of respondents said that they have run stop signs, and 71 percent stated that they never break this rule. Why do the 71 percent who never run stop signs follow the rule? Two people said they don't want to get a ticket (one of these men also expressed this concern with regard to jaywalking). These respondents abide by the rule because of a fear of punishment—getting a ticket. Four respondents cite danger and the potential for accidents as their reason for obeying the law. An additional five people expressed both a fear of being punished (ticketed) and a belief that it is danger-ous to run a stop sign. Only one man said that running a stop sign is "wrong" and he would "feel guilty" if he did it. The belief in the "rightness" of the jaywalking and stop sign rules are roughly equal (and nearly zero), although more people follow the stop sign rule, where the risk of formalized punishment (traffic tickets) is higher or at least is perceived to be higher.

Littering

When asked if they ever litter, one person said that she litters on occasion but tries not to, whereas the remaining 16 people said that they never do. Every person, including the one who said she tries not to litter, expressed a belief in the wrongness of littering as the only reason for not doing it. Whether these people truly never litter is not as important as the fact that they have a belief that littering is wrong and that their goal of not littering is motivated by this belief. People have strong beliefs about this rule, expressing sentiments such as "litter-ing is disgusting" or "wrong," whereas no one said such things about running stop signs or jaywalking. Also, a number of people cited concerns about protecting the environment, saying that a person shouldn't litter if he cares about the environment, and that "God made the world, and we should protect it." Concern with the good of society was also expressed, with one man saying that "littering is detrimental to society" and another saying that it is "socially not right." Compatible with their beliefs, rather than a fear of punishment, some said that they would feel guilty if they littered. Guilt is associated with doing something that the actor perceives as wrong, rather than a response to other people thinking that what he did was wrong. Compliance with the no littering rule is extremely high, as is the belief that littering is wrong. The risk of punishment is perceived as very low. Chaldeans don't litter because they believe it would be wrong to do otherwise. Many appear to have internalized this norm.

Figure 7.6 summarizes the data from this investigation. The black column depicts the overall reported rate of compliance for each of the three public goods. The white and gray columns show the percent of informant explanations that indicated either a fear of punishment (formal or informal) or an internal-ized norm, respectively. For "not jaywalking" and "stopping at stop signs," the white and gray columns can be seen as representing the fraction giving a particular reason, given compliance. For example, of the 71 percent who comply with the law on stopping at stop signs, nearly 60 percent of them indicated a fear of punishment as a motivation. This is not the case with littering, because the one person who reported sometimes littering indicated that it was an error (a personal failing) and that she believes littering is wrong.

These three situations suggest that for any particular public good, people may be motivated by a fear of punishment and/or an internalized belief in the "rightness" of the norm. Furthermore, the same individual may be moti-vated by a fear of punishment in one context, but by his internalized beliefs in another context. Everyone who said that he did not run stop signs because he feared punishment also said that he did not litter because it was "wrong" or

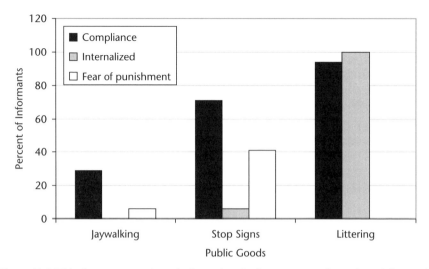

Figure 7.6 This figure summarizes the interview findings on compliance in public goods situations and reasons for compliance ($n = 17$).

"disgusting" (inference: cooperating without the threat of punishment is contextually dependent). Although fears about punishment can substantially increase rates of cooperation, the highest rate of compliance occurred when the respondents believed most strongly in the rightness of the rule or norm—96 percent of people said they never litter, all the respondents believe littering is wrong, yet *none* of them expressed a fear of being punished for littering. However, comparing the fear of punishment and compliance for jaywalking to running stop signs does show that when internalized beliefs are not strong or universal, people still do respond to the threat of punishment and increase compliance. More people stop at stop signs than cross on a red light, and this higher rate of compliance results from a greater fear of punishment in the case of running a stop sign. These data suggest that culturally transmitted norms and beliefs matter more than laws for compliance—all three behaviors in the hypothetical scenarios are against the law, but compliance varies strongly according to whether the behaviors have been moralized and internalized.

Closing

The laboratory work summarized above confirms several of the predictions derived from our social norms models. Experiments show that individuals will administer costly punishment, even in one-shot interactions without reputational possibilities (prediction *a*, from earlier in this chapter), and that the willingness to punish will not depend on the presence of repeated interactions (prediction *b*), but will increase according to the degree of deviation from the norm (prediction *c*). Cross-cultural experiments confirm that the strength of people's willingness to punish varies across societies (prediction *a*) and that some people will be altruistic, even if no chance of being punished exists (prediction *g*). Public Goods Games confirm that people will readily respond to the possibilities of punishment by cooperating more, even before acquiring experience with punishment. Whether adding the possibility of punishment drives the group to full cooperation depends on cultural norms (prediction *f*).

By establishing links of consistency with both experimental results and our theoretical expectations, this ethnographic research adds an important pillar to the foundation of social norms models. In this light, three key points are worthy of emphasis. First, among Chaldeans, punishment for norm violations is fueled by gossip, which allows those best able to effectively administer it to do the punishing (predictions *d* and *e*). Second, punishment is not restricted to cooperative behaviors, and people can be punished for violating a wide range of norms (prediction *a*). Third, people internalize and moralize some behaviors that promote public goods, but not all (or even most) behaviors associated with public goods are moralized. Thus, the culturally transmitted nature of norms means that individual decisions about how to respond will depend on more than just the raw costs and benefits (including punishments) in the situation. It depends on whether individuals have acquired and internalized a norm about what to do in a specific situation, and to what degree they believe others have internalized it. This carries along with it that individuals cannot sensibly be understood as "cooperators" or "defectors" but rather as individuals who have internalized a norm to varying degrees (prediction *a* and *g*) and who will cooperate in some situations but not others.

In the next chapter, we will continue our discussion of social norms models by analyzing two field experiments that were done with the Chaldeans. As you will see, these experimental games demonstrate some of the same patterns observed ethnographically. Combined with post-game interviews, these data confirm that social norms are context specific (prediction *a*), and people (and now specifically Chaldeans) can be altruistic even in the absence of punishment or any chance of repeated interaction or reputation effects (prediction *g*).

8

Culturally Evolved Social Norms Lead to Context-Specific Cooperation

I would never agree to let a partner get more than me. Maybe if my brother was a business partner and needed some extra money the year he was getting married, I would allow him to take more money. But he would not be entitled to any extra money, and any inequality would only be permitted out of the goodness of my heart to help out my brother.

—27-year-old male grocer, explaining the importance of equality between partners

You give beggars on the street $3 or $4.

—51-year-old male lawyer, explaining why he wouldn't accept an Ultimatum Game offer of less than $5

The other person may need the money.

—40-year-old male lawyer, explaining why he would accept an Ultimatum Game offer of $1

Independent of the costs and benefits of a social interaction, do some contexts elicit more cooperation or equity than others? Do certain types of public goods systematically get more support than others? At the root of these questions is a more fundamental issue: Do people have a psychology geared to strictly analyze the costs, benefits, and expected number of future interactions, or are some forms of cooperation and prosocial behavior (i.e., punishment) influenced by the social context rather than just the costs and benefits? Consistent with the emergence of norms developed by the culture-gene coevolutionary process described in the last chapter, mounting evidence suggests that the norms of a group influence the domains in which people cooperate. This is not to say that costs and benefits and evolved human preferences don't influence the norms that govern when it's considered appropriate to cooperate. For example, people place more value on the present than the future (i.e., people have hyperbolic discount functions-this means that people prefer to have things now rather than later and that they're willing to pay more to get things sooner; Camerer 1995). So we suspect that public goods with immediate benefits will be favored over those with only long-term payoffs. We argue that cultural norms, evolved human preferences, and costs and benefits are all important factors in explaining cooperation.

The theoretical ideas developed in chapters 2, 3, and 7 suggest that cultural evolutionary processes can create different amounts of punishment and cooperation in different domains of social life. This implies two things. First, within cultural groups, people may be very prosocial in one domain and quite noncooperative in a different domain, even when the costs and benefits are similar. Second, different cultural groups may cooperate and punish to substantially different degrees in a domain with very similar costs and benefits. Our own ethnographic experience among the Machiguenga in the Peruvian Amazon and the Yasawans in Fiji reveals a stark contrast in this regard. Both the Fijians and Machiguenga live in small villages and rely principally on horticultural production (mostly of root crops). Both groups could benefit from cooperation and cooperative maintenance of village public spaces. We and other ethnographers (Baksh 1984; Johnson 2003; N. Smith 2001) have repeatedly documented the failure of cooperative labor and public projects among the Machiguenga. Machiguenga leaders, sometimes inspired by missionaries, announce public projects and cooperative gardening ventures. When the day comes, few people show up, and those who come do little and depart early. Multiple times, we've observed the village leader blowing his horn repeatedly over the course of the day to announce a village meeting or activity, but few people respond. In one case, the students—forced by their teachers—had to build their own school because adult males would not show up for cooperative labor. In contrast, Fijian cooperative labor, both gardening and school maintenance, happens like clockwork. Announcements are made in a ritualized style (by a person whose role is assigned by heredity descent), and people respond. Although Fijians are rarely on time, people always show up eventually and generally work hard, always laughing and joking with each other as the work proceeds. The difference between Machiguenga and Fijian villages in solving public goods problems could not be starker. Fijians succeed because they have culturally inherited a vast number of beliefs and practices, ranging from the belief in the hereditary decision-making authority of village chiefs to beliefs in the supernatural punishment of those who oppose the collective will. Machiguenga, on the other hand, arrive in adulthood with none of this, but instead have a sense of self-reliance and rugged individualism that would make even an American cowboy feel like a dependent conformist.

Although Fijians can effectively solve many public goods problems, they are not cooperative in every domain. We observed numerous instances in which a person was assigned the job of collecting and protecting sums of cash, intended for use in a specific village project (e.g., fixing the school or buying a boat engine). Inevitably, when decisions were finally made to use the cash, it often turned out that the money had been substantially dipped into by the holder, who always had a good reason and promised effusively to pay it back (and rarely did). Or sometimes, when the money is called for by the village elders, disagreements about the accounting arise, with the money's holder claiming there was supposed to be less than what others believed there ought to be.

Ethnographically, the domain specificity of cooperative behavior among the Hadza was related to us independently by two different Hadza researchers (Frank Marlow and Nicholas Blurton Jones). The Hadza are nomadic hunter-gatherers in Tanzania and are renowned for their meat sharing. Game, especially big game, is shared among all members of a hunter's camp, and sometimes with people from other camps. Meat sharing is expected, and people are ready to punish nonsharing. Thus things usually go smoothly. Interestingly, however, when anthropologists tried to pay a group of Hadza men (who routinely share meat) with a lump sum of tobacco for their assistance, the men begged the anthropologists to give out individual portions. The men explained that if they had to divide it among themselves, fights would break out, and someone could end up dead. The anthropologists divided it equally and handed out individual portions. Clearly, the "rules" for cooperative meat sharing don't extend to cooperative tobacco sharing. Quite systematic work on cooperation among the Aché of Paraguay further confirms the context specificity of sharing and cooperation among another group of foragers (Gurven et al. 2002; Gurven 2001, 2004).

In this chapter, we use a combination of experimental and interview data from the Chaldeans to show that cultural norms influence prosociality and create context-dependent cooperative behavior.

Ultimatum and Dictator Games

Experiments Reveal That Cultural Domains Influence Cooperation

To augment our ethnographic inquiry among the Chaldeans, we administered two economic experiments, the Ultimatum Game (UG) and the Dictator Game (DG). The experimental data gathered from these games allow us to study variation in prosocial behavior among Chaldeans and to compare Chaldean behavior to other groups, especially "typical" American subjects. Further, when combined with our in-depth interviews and our detailed understanding of the ethnographic and historical context, the experimental results allow us to examine how local cultural context influences decisions made in the game. In a sense, it helps us figure out what the game is actually measuring.

As noted earlier, the UG looks at both individuals' willingness to punish others who are unfair to them (second-party punishment), and people's response to the threat of punishment. Complementing the UG, the DG measures people's altruism or fairness in the absence of any threat of punishment. One might theorize that adding the threat of punishment should increase cooperative behavior by adding the threat on top of any existing taste for altruism (e.g., Fehr and Schmidt 1999; Charness and Rabin 2002; Bolton and Ockenfels 1999). However, Chaldeans seemed to understand the two games in quite different ways, such that they offered more (were more fair-minded and cooperative) in the absence of a punishment threat. Interestingly, in the Chaldean case, people actually showed more fairness and cooperative behavior and less self-interest in the absence of punishment. These results coincide with norms for appropriate behavior in two contexts that are important in Chaldean life: business and charity. That is, Chaldeans tended to perceive the UG as a businesslike competition and the DG as an opportunity for charity toward (or sharing with) other Chaldeans. Below, we explain the UG and the DG (describing the experiments, methods, and results) and then illustrate how the decisions made in the experiments are consistent with Chaldeans' real-life behavior and reflect domain-specific norms for social behavior.[1]

The Ultimatum Game

As explained in chapter 6, the UG is a bargaining experiment involving two players who are anonymous to each other. The first player, called the proposer, is given a sum of money and told to offer some portion of it (from 0 to 100 percent) to the second player, often labeled the responder. The responder can either accept or reject the offer. If she accepts it, she gets the amount of the offer, and the proposer keeps the remainder of the money. If she rejects it, neither player receives anything. For example, imagine that the proposer is given $100, and he offers $35 to the responder. If the responder accepts the offer, he gets $35 (the amount of the offer) and the proposer gets $65 ($100 less the offer of $35). If the responder rejects the offer, both the responder and the proposer get $0. If people are rational and self-interested, proposers should offer the smallest non-zero amount possible, and the responder should always accept any positive offer. The reasoning goes as follows: Getting something is better than getting nothing. Consequently, a responder should accept an offer of $1 because if he accepts it he gets $1, but if he rejects it he gets $0, and $1 is better than nothing. Knowing this, the proposer should offer $1.

The UG has been tested in more diverse cultural settings than any other experiment and is probably second overall as the "most run" experiment, perhaps surpassed only by the Prisoner's Dilemma. In addition to the "standard subjects," who are made up almost exclusively of American and European university students, the game has been played with students in Indonesia, China, Japan, and Israel (Camerer 2003; Roth et al. 1991), and with 15 small-scale societies scattered across Africa, South America, New Guinea, Indonesia, Oceania, and Asia (Henrich et al. 2004). Four general observations can be made about the results. First, UG behavior is extremely homogeneous among adults from industrialized societies—more recent work has included both students and nonstudents (Carpenter et al. 2005). Among these societies, proposers make mean offers in a range from 36 percent to 48 percent, with modal offers at 50 percent. Rejection rates are more variable, with people in different experiments rejecting from 4 percent (U.S., Forsythe et al. 1994) to 33 percent (Jerusalem, Roth et al. 1991) of offers, and frequently rejecting offers below 20 percent. Second, UG behavior that includes subjects from small-scale societies shows much more heterogeneity, both within and across groups. In the 15 small-scale societies, mean offers ranged from 26 percent to 58 percent, with some groups not rejecting any offers (even very low offers) and other groups rejecting offers of more than 50 percent (Tracer 2003). Though some groups have mean offers as low as 25 percent, nowhere do people approach the self-interested game theoretic solution. Third, these substantial cultural differences seem to reflect differences in the daily life of people in these societies. Subsequent cross-cultural research has replicated and extended these findings (Henrich et al. 2006).

How might we interpret such findings, given all that we've learned? We think they show that (1) different groups have different notions of fairness, (2) different groups have different proclivities for punishment (rejecting an offer is a form of costly punishment—the responder incurs the cost of not taking the offer to prevent the proposer from getting any money), and (3) the "right" behavior in the game is influenced by what people learned from growing up in a particular place. Students typically offer close to 50 percent, from which we can conclude that students share a belief that in this type of situation, both players should benefit equally (or close to equally) from participating in the game. Whether or not the individual proposer believes in the value of fairness, the group as a whole has this norm—and as we saw in our interview data in the last chapter, people may adhere to a norm because they believe it is right or because they fear punishment. In the UG, this works itself out in two ways. If the proposer doesn't share the fairness norm but still offers near 50 percent, then most likely he fears that a lower offer would be rejected. Why would he fear rejection of his offer? Because he knows that at least some other people in the group believe in fairness (that is, they think 50/50 is the "right offer") and would perceive a lower offer as a punishable norm violation. Second, if the proposer does share the fairness norm, then he would offer near 50 percent, and the threat of punishment would not matter to him because he would make an equitable offer even without the possibility of rejection.

Using the DG, we can begin to separate out people who make fair offers because they believe in equity from those who do it out of a fear of punishment. The DG is very similar to the UG except that now the responder cannot reject. That is, the proposer—now called the dictator—is given a sum of money and can allocate a portion of this money (from 0 to 100 percent) to the responder, now called the recipient. The recipient receives the money she is allocated, and the dictator keeps the remainder of the money. Since the proposer, now a dictator, does not have to worry about her offer being rejected, the amount that she allocates to the recipient should reflect her belief in what is an appropriate division. The DG allows us to peek into the mind of the dictator to see if she has a preference for fairness/equity (in this context) and, on aggregate, to assess the extent to which fair behavior is driven by a belief in fairness as opposed to acting fairly out of a fear of punishment.

When proposers make low offers, there is a different explanation for the behavior than when proposers make equitable offers. As we already discussed, equitable offers are an indication that either the proposer believes in equity or knows that other people believe in equity and will punish people who make inequitable offers. But when proposers make low offers, it can be because (1) the proposer believes it's OK to make low offers and other people also feel that way, so there is no fear of punishment; or (2) the proposer believes it's OK to make low offers, and other people think it's wrong, but they don't have a norm for punishing in this context. With both of these explanations, fear of punishment is not a factor in determining the proposer's behavior.

Ultimatum Game with the Chaldeans

Methodology

With few exceptions, Natalie played the UG with people whom she had just finished interviewing. In contrast to experiments done at universities where the subjects don't know the experimenter, Natalie usually spent about 1.5 hours interviewing each subject, during which time she collected data about their lives, before she played the game. Following this interview, the experiment was presented as something separate in which they could choose or decline to participate.[2] Combining experimental data with interview data and ethnographic observation enables us to interpret the results in context and with an understanding of daily life.

The UG was conducted with 30 pairs of players, ranging in age from 19 to 70, with a mean age of 34.5. Fifty-four percent of the players were male. The game was played for a stake of $20.[3] The game was fully explained to both proposers and responders, and the players were told that they were playing with another Chaldean from their community but that the other person would remain anonymous—it was emphasized that neither player would ever know the identity of the other. The game was explained by moving 20 $1 bills around to visually demonstrate the game. Before playing, each subject had to correctly answer a test question to verify that he or she understood. Players were not told their role until after the game was completely explained and play was about to commence. If the person was a proposer, she would tell Natalie the amount of her offer and Natalie would let her know if it had been accepted or rejected within a week. During that week her offer would be taken to another player, who was assigned the role of responder. If the responder accepted the offer, he was paid immediately and Natalie would pay the proposer (either in person, or by mail). If the responder rejected the offer, she contacted the proposer (either in person, or by mail) to notify her of the rejection. To gain additional information beyond that obtained from the experiment, an extensive postgame interview was conducted with each player.

UG Results

Figure 8.1 shows the distribution of Chaldean offers, and the acceptances and rejections of those offers. The overall height of the bars (white plus black) shows the frequency of offers at each offer level. The white portion of the bar shows the

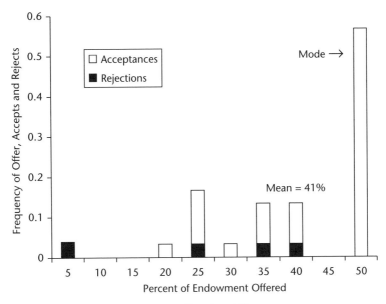

Figure 8.1 Chaldean Ultimatum Game offers ($n = 30$).

frequency of offers at each offer level that were accepted. The black portion of each bar shows the fraction of offers at each offer level that were rejected. Chaldean proposer's made a mean offer of 41 percent, with a mode of 50 percent. The mean offer from this sample of Chaldeans is somewhat lower than that typically found among both American university students and American adults (Camerer 2003; Cameron 1999; Carpenter, Burks, and Verhoogen 2005; Forsythe et al. 1994; Henrich 2000; Hoffman et al. 1994; Roth et al. 1991). In order to test whether the Chaldean offers were significantly lower than those from other subjects, we compared these offers to offers made by Forsythe et al.'s (1994) students and Ensminger's (n.d.) adults from rural Missouri. Using the Mann-Whitney (MW) test to compare means of non-parametric distributions, the Chaldean offers are likely different from those of the students ($p = 0.07$) and Ensminger's adults ($p = 0.04$). All three distributions are shown in figure 8.2.

Interpreting these relatively small differences among Chaldeans, students, and rural Missourians is a bit tricky for several reasons. First, although the UG is fairly robust among subjects from industrialized societies with regard to minor changes in the protocol, variations in the instructions and presentation can produce small differences between identical subject pools. Second, an emerging finding is that undergraduates aren't "finished being socialized." Developmental comparisons suggest that age doesn't matter once subjects are older than about age 24, but up until then, prosocial behavior continues to increase with age (Carpenter et al. 2005; Harbaugh and Krause 2000; Harbaugh, Krause, and Liday 2002). Thus, the differences between both the Chaldean and Missourians, compared to students, likely results from developmental differences (captured by the difference in average ages). However, the

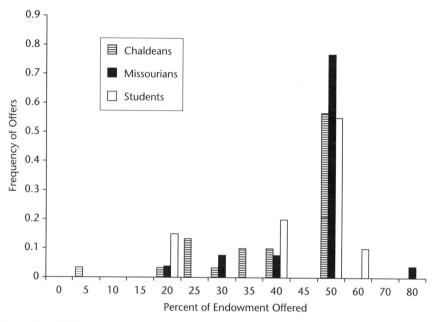

Figure 8.2 Ultimatum Game offers made by Chaldeans, Missourians (Ensminger n.d.), and students (Forsythe et al. 1994). Rejections are not shown. Chaldean vs. students: $p = 0.07$; Chaldeans vs. Missourians: $p = 0.04$ (Mann-Whitney).

larger difference between Chaldeans and Missourians does not likely result from age, income, or sex differences between the samples (as these don't matter within the samples, and all three are consistently nonpredictive of UG behavior across many samples). Therefore, we suspect this may capture a cultural difference. Other cultural differences have been noted within the United States (Cummings and Ferraro 2002).

Complementing this somewhat lower mean offer, some Chaldeans are also willing to reject inequitable offers as high as 40 percent (fig. 8.1). Using the rejection data, it is interesting to ask the question of what a proposer would do if he both (1) wanted to make as much money as possible and (2) knew the likelihood of rejection at each offer level. We call this offer the "Income Max-imizing Offer" (IMO). By deploying a method that is beyond the scope of this book (Henrich, et al. 2004: appendix), it is possible to calculate the likelihood of each offer-level being accepted with respect to the rejection pattern of these subjects. In a community in which nobody ever rejects any positive offer, the IMO is the smallest possible amount. In contrast, in a community in which all inequitable offers are rejected, the income maximizing strategy is to offer 50 percent (among the Missourians, for example, IMO = 50 percent). Given the pattern of Chaldean rejections, an income-maximizing Chaldean should offer 40 percent. The offer data show a mean offer of 41 percent and a modal offer of 50 percent, so Chaldean proposers are pretty close to income maximizing, with some erring on the side of caution.

The convergence of the IMO and actual offers suggests that proposers and responders have the same expectations about what is appropriate behavior in the game, and that proposers were able to adjust their offers to match the responders. Although the responders do not behave in a way that is consistent with purely self-interested motives (self-interested responders should accept any non-zero offer), Chaldean proposers' behavior is generally consistent with selfish motives, as they seem to accurately perceive the likelihood of the low offers being rejected and manage to come pretty close to maximizing their own payoffs (on average). Consistent with many other U.S. and European findings (Fehr and Schmidt 1999), responders seem motivated by a concern with inequity, whereas proposers' behavior is explicable principally in terms of self-interest. Although proposer behavior is consistent with self-interest and maximizing income, it is also consistent with a belief that responders should get a significant share of the money. The UG, on its own, doesn't allow for a determination of which is the actual motivator.

In analyzing the variation in offers, multivariate regressions show that no demographic variables explain the UG offers. Age, sex, number of years in United States, age when immigrated, income, and household size all lack any significant relationship with offers. That is, proposer offers don't systematically vary with these measures. This is a standard result found among subjects from many countries and cultures (Cameron 1999; Fehr, et al. 2002; Gowdy, Iorgulescu, and Onyeiwu 2003; Henrich et al. 2004). As well, in our data, a Chaldean's degree of ethnic identity, cooperation with Chaldeans, and cooperation with non-Chaldeans also have no predictive ability to explain offers (these variables were measured using ethnographically derived indices; see chap. 9). The lack of predictive power is not surprising given the low degree of variation in UG offers (standard deviation = 0.12).

The Dictator Game

Until recently, the Dictator Game had been conducted almost exclusively with students in industrialized developed countries. Some exceptions have used, as players, adults in rural and urban Missouri (Ensminger n.d.), employees at a midwestern distribution center (Carpenter et al. 2002), and people living in two small-scale societies in Africa (Marlowe 2001; Ensminger 2001) and in one small-scale society in lowland South America (Gurven 2001). In U.S. student populations, under conditions of player anonymity, DG allocations to the second player drop dramatically in comparison to offers made in the UG. In fact, in some experiments (see fig. 8.3), the DG distribution is a near mirror image of the UG distribution with a single mode at 0 percent (Forsythe et al. 1994: low stakes condition), instead of 50 percent.

This dramatic decline in money given to the recipient strongly suggests that a substantial share of students act equitably in the UG because they believe a lower offer would be rejected. In comparison, Dictator Games with adults in the

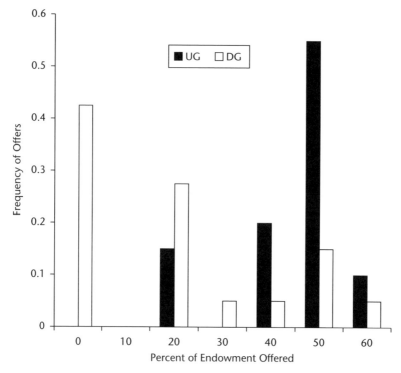

Figure 8.3 Typical Ultimatum vs. Dictator Game offers made by U.S. students in low-stakes experiments (Forsythe et al. 1994).

United States (Carpenter et. al. 2002; Ensminger n.d.) indicate that fairness matters more than punishment in choosing an offer, as these subjects offered virtually the same in the UG and DG even though the DG is a punishment-free situation. The most obvious explanation for this is, as described earlier, that university students aren't finished growing up yet and have not fully acquired their adult social motivations (this leaves one wondering why experimentalists interested in adults continue to concentrate their research efforts on students).

Dictator Game with the Chaldeans: Methodology

As with the UG, Natalie conducted the DG after an interview; in this case, it was their second interview. Since the second player could not reject the money allotted to him, all subjects were paid immediately—as soon as the dictator told Natalie how much he was giving to the other player, he was paid the remainder. As with the UG, players were told that the other player was a Chaldean from their community, but his identity would remain anonymous. In this game there were 21 dictators, and they played with a stake of $20. Some of the same subjects who played the UG also played the DG; this was done so that we could compare results from the same people across the two games.[4] In order to minimize any effect of one experiment on the other, 3 to 9 months elapsed between the times that the subject played the UG and the DG.[5] Players ranged in age from 21 to 65, with a mean age of 36.8. Thirty-eight percent of the players were males.

Chaldean DG Results

Chaldean DG offers not only remained as high as in the UG (as in previous research with nonstudents), but they increased. On average, dictators gave 44 percent to the recipient (as compared to 41 percent in the UG), with 81 percent of the dictators giving the modal offer of 50 percent. Figure 8.4 shows the distributions of UG and DG offers. Included in the DG offers is an offer of 0 percent that, on the surface, appears to be a completely self-interested act. However, during the game this individual said that he wanted to give all the money to the other player, but that first he wanted the other person's name. He gave the impression that he planned to reveal his own identity to the other player in order to get recognition for his generous act—after all, he was going to give everything to the other player. When Natalie declined to divulge the name, the player said that he would take everything for himself but then give all of it to someone who needs it. We don't know if he really gave the money away, but this apparently selfish offer may actually have been hypergenerous but only under the condition of nonanonymity. If we remove this offer from the data, the mean DG offer rises to 46 percent and the difference between the mean UG and DG offers increases to $p = 0.07$ (MW).[6]

The extremely equitable offers in the DG appear to be an anomaly when compared to the robust results from experiments with students (e.g., Forsythe

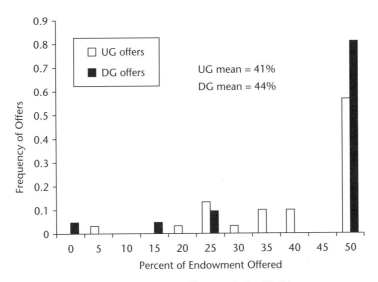

Figure 8.4 Ultimatum vs. Dictator Game offers made by Chaldeans.

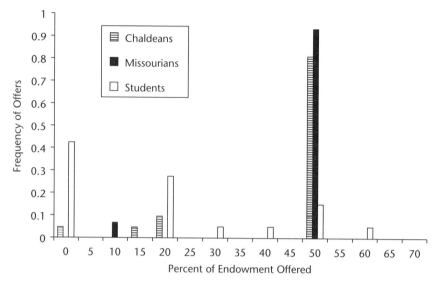

Figure 8.5 Dictator Game offers made by Chaldeans, rural Missourians (Ensminger n.d.) and students (Forsythe et al. 1994). Chaldeans vs. Missourians: $p = 0.32$; Chaldeans vs. students: $p < 0.00$ (Mann-Whitney).

et al. 1994; Hoffman et al. 1994). However, as with the UG, the Chaldean offers are consistent with DG offers made by non-student adult subjects. Examples of average DG offers among adults are 47.3 percent (rural Missouri), 48 percent (urban Missouri), 45.4 percent (employees at a midwestern distribution center: Carpenter et al. 2002), and 42 percent (Igbo villagers in rural Nigeria: Gowdy et al. 2003). What makes the Chaldean offers potentially unique is that their DG offers were higher (marginally significant) than their UG offers, in contrast to the indistinguishable UG and DG offer distributions of these other nonstudent samples. Figure 8.5 compares the distributions from three of these groups.

Multivariate regression analysis shows that demographic variables including age, sex, income, number of years in the U.S., age of arrival in the U.S., and household size do not predict DG offers among the Chaldeans. This is consistent with similar negative findings with other DG data (Gowdy, et al. 2003). As with the UG, subjects' degree of ethnic identity and cooperation with Chaldeans and non-Chaldeans do not have any significant relationship with offers, and the lack of predictability of these variables is strongly affected by the lack of variation in the offers (standard deviation = 0.14).

Postgame Interviews Support the Results

The mean offer made by Chaldean players increased from the UG to the DG (note that the UG was played first, so if there was a "learning effect," it drove offers up). Despite the change in mean offers, the modal offer (50 percent) remained constant in the two experiments. These results raise two questions:

What caused the mean offer to increase? And did the people who offered 50 percent in both games do so for the same reason in each experiment, or did their motivations differ in the two games? Evidence from postgame interviews and ethnographic work reveals that players perceived the DG and UG quite differently.[7] The DG was not perceived as just a UG without the possibility of punishment, but as an entirely new situation. Each situation seemed to evoke different goals and preferences.

Different Contexts, Different Motives

Removing the punishment possibility did more than change the players' cost-benefit calculation; it changed how they comprehended the artificial game situation and how they felt toward the other person. In the UG, the possibility of having one's offer rejected made the game competitive, which appeared to cue rules, models and/or preferences that underlie behaviors like those used in business interactions. As you will remember, Chaldeans often compete fiercely with other Chaldeans in business. This explains why some proposers shaved the line so close to the IMO. In contrast, the second player in the DG has no means of retribution and is at the dictator's mercy. This circumstance evoked a sense of responsibility to help their partner—feelings of competition were absent from this context. Norms governing behavior in charitable situations were cued in the DG. The "Business Competition Game" became the "Charity Game" (we did not actually use these labels in the game). Recall from chapter 7 that Chaldeans are quite serious about giving charity to other Chaldeans (remember the outrage expressed about the rich man who donated nothing to help Chaldeans in Iraq).

This shift in people's interpretation of the contexts became evident in post-game interviews (Table 8.1). Numerous players talked about their reasons for their choices differently in the two experiments, describing their UG decisions in terms of business deals (a competitive context) and their DG decisions in terms of charity, helping, and fairness (all noncompetitive sentiments). For example, a UG proposer explained that she offered 40 percent in the UG because it is her "negotiating strategy; never give half...always try to get a little more than what's right down the middle." When people negotiate they are competing with each other to try and get the "best deal," and it seems that she thought about the UG in terms of the two players each trying to get the most for herself. But after offering 50 percent in the DG, she talked about wanting the game to "be even" and explained that she wasn't losing anything by giving money away because she's still gaining from having participated. Similarly, an investment broker who accepted an offer of 40 percent in the UG said that he doesn't understand why anyone would ever reject an offer in the UG, and then added, "It must be my 'investment thinking.'" He was viewing the game the same way that he would think about making investments, where the whole point is to make money, and something is always better than nothing. However, in the DG, he offered 50 percent even though there was no chance of rejection. In fact, after making the DG offer, he expressed sentiments of fairness, explaining that he wanted to keep

Table 8.1 Summary of Proposers' and Responders' Perception of Context*

Explanation	UG	DG
Fairness	8	14
Charity	0	3
Competition & self-interest	8	1

* The total number of responses is less than the total number of players because some interviews did not yield any relevant information.

some of the money but that the other player should get some also. Where did his "investment thinking" go during the DG? A more dramatic shift took place in a man who offered 25 percent in the UG and 50 percent in the DG. With regard to his offer in the UG, he said "It's business." In fact, he said he was going to offer 5 percent ($1) but decided at the last minute to give a little more as a "consolation prize for playing." But after the DG, he talked about his offer in terms reminiscent of charity, saying that the other player may be poorer than him and may need the money, and that perhaps he should have given more than half. He certainly didn't seem concerned about the relative wealth and need of the responder in the UG. Why suddenly this concern with the other player's well-being? Table 8.1 summarizes responses from our post-game interviews by classifying responses into categories for fairness, charity, and competition/self-interest. For these players, their postgame comments suggest a change in the framing of the games from one of competition in the UG to one of camaraderie or charity in the DG, and with this contextual shift came a change in behavior such that subjects displayed culturally-appropriate behavior for the shifted context.

Evidence for the contextual variation also emerges among players who offered 50 percent in both experiments. Although their behavior remained constant, the motivation for the offers varied as the perceived context changed. A UG responder who accepted 25 percent said that had she been a proposer she would have offered 50 percent because the responder may be greedy and reject anything less than half—in the UG, fear of rejection would have motivated her equitable offer. However, after offering 50 percent in the DG, she said that she made this offer because "it's fair." In the UG, she didn't express any concern with fairness, only with the possible greed of other people. Another player displayed a similar shift in motivation. This woman, during the explanation of the UG, seemed focused on the possibility of low offers' getting rejected. At one point, Natalie did an example of an offer of 10 percent, for which she showed how much each player would receive if the offer was accepted and how much each would get if it was rejected. The woman interrupted to ask, "No one would take that, right? They'd both end up with nothing." In her mind, it was clear that people would reject low offers. She offered 50 percent. When she played the DG, she again offered 50 percent and this time her decision couldn't have been affected by a concern about rejection. Further illustrating the shift in context, 17 percent of DG proposers who gave postgame explanations for their offers talked about helping the other player or expressed concern about the other player's neediness, whereas no one mentioned this in the UG.

In both experiments, notions of fairness influenced many people, affecting how much was offered by proposers/dictators and whether a responder accepted or rejected his offer. Subjects made comments like, "I like things to be equal," and "I offered $10 because it's fair . . . both people should get the same amount of money because they'll have spent the same amount of time with you [the experimenter]." Jennifer rejected an offer of 40 percent and was outraged at the proposer for making an inequitable offer. With great vehemence she exclaimed, "Of course they want more, but it's free money, and we didn't work for it. . . . Share the wealth! Think of the other person!" She seemed to be saying that it was just luck that the other player got to be the proposer, and consequently he wasn't entitled to a greater share of the money. Since he didn't earn the money, it is only fair to share it equally.[8] Another Chaldean woman concerned with fairness accepted an offer of 50 percent and said that she would have rejected anything less than an even split. Why? "Because," she said, "it's fair. . . . Fair is fair. If you have $20, why the hell would you give me less than $10? I know you're no better than I am." Even though players in both experiments cited fairness as an important factor in their decision making, fairness mattered more to DG proposers than to UG proposers. Of the players who gave postgame explanations for their decisions, 50 percent of UG subjects mentioned fairness in comparison to 66 percent of subjects in the DG.

Ethnographic Evidence Also Supports the Results

Chaldeans appear to have multiple rules or models in their heads about how to behave in different real-life situations, and these promote cooperation or altruism in some contexts and self-interested behavior in others. These context specific situations for cooperation map onto the different behaviors observed in the UG and DG. Ethnographically, we found that Chaldeans were competitive in their business relationships, acting cooperatively by favoring business relations with coethnics but working to get the best individual deals in these relationships (chap. 9). Within these coethnic relationships, for example, clients try to get their lawyers to charge them less than the usual rates while expecting more time and attention for their cases, wholesalers try to convince store owners to increase their orders, and patients ask their doctors to give them free medicines. In each of these cases, the client, wholesaler, and patient each give enough to their partner (the lawyer, store owner, or doctor) that they all agree to stay in the relationship, but each person pushes the envelope to see how much he can get for himself without jeopardizing the partnership.

A similar interaction appears to happen in the UG. The proposer offers less than 50 percent but not so much less that he thinks the responder will reject the offer. Although 57 percent of Chaldean proposers offered an equal share to the responder, 43 percent tried to get at least a little more for himself or herself. The UG likely triggered a "business model" for the Chaldeans. Business relationships are perceived as interactions within a competitive dyad (e.g., grocer versus supplier, lawyer/financial advisor versus client), in which either partner can terminate the relationship if the other acts too unfairly. Even though the people involved need each other and cooperate to an extent that allows the relationship to persist, each person is in the relationship to satisfy his own needs and do as well as he can for himself. These characteristics can easily be superimposed onto the UG, making it the framework in which the game is frequently played. In the UG, the responder's ability to reject the proposer's offer made the relationship between the players appear competitive, so that it resembled a business relationship. And as in business, the proposer tried to get as much as he could for himself while recognizing that his partner would terminate the relationship (i.e., reject the offer) if the distribution was too unbalanced.

In contrast to the UG, the DG seems to have cued a charity/assistance model for Chaldeans. Simply removing the competitive element (removing the responder's ability to reject) in the DG appears to have triggered a different set of interactional rules and norms from Chaldean culture. The importance of helping other Chaldeans is a robust belief in the community and, in particular contexts, this belief is acted on. Common situations for charity are giving money to needy Chaldeans in Iraq and donating money and essentials (e.g., food, furniture) to recent Chaldean immigrants in Detroit. Evidence that Chaldeans value helping others emerged when people were asked about their reaction to a hypothetical situation in which a wealthy Chaldean man failed to contribute to collection at church (chap. 7). People said that they would think the man was cheap, stingy, bad, and that they would think poorly of him, or wonder why he didn't give. These negative reactions imply that the man did something wrong by not giving to charity, thus suggesting a norm or belief that Chaldeans should be charitable to Chaldeans. As well, people frequently spoke with pride about Chaldean generosity and the practice of helping others and being hospitable. Outside of business, Chaldeans take satisfaction and pleasure in giving of themselves and incurring personal costs to help others. Since the DG is only one-shot and anonymous, reputational benefits are clearly not all that is driving their charitable contributions (our interviews indicated that people fully believed the one-shot anonymous nature of the games).[9]

Conclusions

At first glance, the UG and DG results from the Chaldean experiments are puzzling. Most theories predict that offers should decrease in the DG relative to the UG because it is assumed that at least some of the high UG offers occur because of a fear of rejection and that these players should lower their offers in the DG. But when we look at the results in a cultural context and consider the ethnographic evidence, the offers made by the Chaldeans make sense. Chaldeans, like all other humans, have culturally learned, context-specific rules for cooperation and competition. The experimental games, which are presented in neutral language and with no explicit context, can be interpreted by players in a way that fits into their cultural frames and that trigger rules appropriate for that context. The UG, with explicit competition in the form of the responder's ability to reject, appears to have cued a business/competitive model, and players acted accordingly by trying to get a little more for themselves. In contrast, the DG, which lacks explicit competition by turning the responder into a recipient, triggered a robust social model of charity toward fellow Chaldeans that led to an increase in offers compared to those made in the UG.

The postgame interviews show that the DG cues rules about charity and fairness, with players feeling a need to help, or share with, the recipient. The offers in the DG were so high because it cued a context in the minds of the players that has strong social norms for generosity. Along similar lines, Eckel and Grossman (1996) played a DG with students where the recipients were either another student or a charitable organization (Red Cross). On average, offers made to the charity were nearly three times higher than offers to fellow students. Their results add further support to our observation that people, at least in some cultures, cooperate more in charitable domains. It is interesting that the Chaldeans' charity rules were cued simply by removing punishment from the game and that they applied these rules to individuals regardless of whether they were actually in need. In contrast, the students in Eckel and Grossman's study applied charity rules only when the game was explicitly framed in this manner, and the recipient was an actual charitable organization.

We can use evolutionary-based cultural models to help explain the variation in behavior across experiments within a group and the variation across cultures in the same experiments. These models are founded on a premise that all humans have an evolved psychology that limits the range of rules that can be applied in any context.[10] There is variation across experiments within a group as different sets of rules are cued by different games. There is also variation across cultures because the same rules are not always applied in the same contexts as we move from culture to culture. However, our limited number of evolved rules (or mental models) for interaction restricts the degree of variation across contexts and groups. By understanding the subjects' culture, we can figure out which mental models are applied in the different games and, more importantly, which mental models are applied in different real-life situations.

Consistent with our ethnographic and interview data, this field experimental work indicates that both punishment and fairness play roles in cooperation, although not necessarily in the way we might expect. Among Chaldeans, it seems that a belief in fairness is more likely to induce cooperation than is a fear of punishment. The higher offers in the DG correspond with a heightened concern with fairness in this game, relative to the UG. Although a fear of punishment in the UG did drive up offers, they did not reach as high a level as when fairness/charity was driving them. Punishment can boost cooperation, but a belief in fairness (or that cooperation is the right thing to do) may be at its foundation, just as it is with regard to littering, jaywalking, and running stop signs (chap. 7). Part of the reason for variation in cooperation across contexts may be that fairness is not expected, or considered appropriate, in all situations, and that levels of cooperation may correspond to context-specific, culturally influenced beliefs and expectations for fairness.

9

Ethnicity
In-Group Preferences and Cooperation

Being Chaldean isn't about the church. It's about being accepted.

—24-year-old male consultant

Chaldean is a race. You're born Chaldean. . . . You can't acquire it later or give it up.

—40-year-old male lawyer

As sketched in chapter 3, evolutionary models examining the dual inheritance of culture and genes indicate that the human motivation to interact with, and learn from, members of one's own ethnic group are rooted in the benefits accrued from interacting with people who tend to share your norms and expectations (Gil-White 1999, 2001; McElreath, Boyd, and Richerson 2003).[1] Social and economic interactions occur more smoothly when the people interacting are playing by the same rules. When all people involved agree on the right way to interact, there is a greater chance of all parties coordinating their interaction and avoiding punishment. Often, members of a cultural group share the same interactional norms among themselves and possess interactional norms different from other groups. Although some social norms may be common to multiple groups, more norms will be common within groups than among groups. Consequently, the evolutionary models show that a person who chooses to both preferentially interact with and learn social behavior from other members of his own group will be more likely to have positive interactional experiences (and get higher payoffs, on average) than a person who does not. In this way, cultural evolution spontaneously generates a social environment that favors both genes and cultural traits for an ethnic psychology.

To gain a deeper understanding of this idea, let's consider the example of marriage norms. Many cultural groups have rules about payments for marriage—either the husband and his family pay the bride's family (a bride-price) or the wife's family pays the husband's family (a dowry). If a couple from a bride-price society planned to get married, the husband's family expects to pay, and the wife's family expects to be paid. However, if a man from a dowry society wants to marry a woman from a bride-price society, then there could be serious problems, as each would expect to receive a payment, and neither would expect to pay. Clearly, a couple from the same payment system will have a much easier time negotiating their marriage-related financial obligations than will a couple from different systems. The issue is not whether bride-price or dowry marriage systems are better or worse, but rather that success in many social interactions hinges on coordinating behavior and behavioral expectations.

Important differences between individuals in their preferences (e.g., for child-rearing practices), ways of making decisions, styles of interactions, and expectations of others (responsibilities of wives, children, or business partners) often represent hidden properties that surface only after an interaction between individuals is well under way, and these interactions are often costly to disengage from. If there existed an easily identifiable marker of who is likely to possess compatible norms, then individuals who make use of this marker would prosper relative to those who do not. This marker would allow individuals to figure out both (1) who to interact with to avoid costly miscoordination, and (2) who to pay attention to for learning the norms that are most likely to coordinate with those in one's social environment.

The nature of cultural transmission (e.g., prestige and conformity biases) creates just such a circumstance: certain markers will tend to be associated with underlying social norms. In many circumstances, these markers are sensibly called *ethnic markers*, although they may or may not correspond to common or popular divisions of ethnic communities. For an ethnic marker to be useful in identifying members of one's group, the markers must be readily evident and difficult to fake. Language and accent provide excellent examples. Americans who try to fake a British, Irish, or Scottish accent may be able to fool their fellow countrymen, but few can fool native speakers for very long. For most individuals, languages or dialects must be learned by around age 13 in order to speak with a native accent. This means that if you hear someone speaking English with what you perceive as an accent, they probably did not grow up where you did. Of course, alongside such hard-to-fake cues as accents and mannerisms are dress, hairstyles, greetings, tattoos, food taboos, arrowhead styles, and other adornments. Together, these provide important cues that individuals can use to figure out both who to pay attention to for learning less obvious social norms and who to preferentially interact with to maximize the chances of beneficial coordination.

By producing a reliable association (covariation) between such markers and underlying norms, cultural evolution creates a regularity that genetic evolution can exploit. Following others, we hypothesize that natural selection acting on genes favored a psychology to preferentially learn from, and interact with, individuals who share your culturally transmitted markers. For example, learning the details of bride-price or dowry from someone who shares your accent (or from your point of view, has no accent) or preferring to trade with someone who dresses like you and speaks your language is part of our species' psychology.

One interesting aspect of this ethnic psychology is that people tend to essentialize members of ethnic groups by thinking of their ethnicity as part of their underlying essential nature. Gil-White (2001) proposes that people conceive of ethnic membership as arising from an immutable essence that a person is born with.[2] Humans are biased to believe that although individuals can change their ethnic markers, they can't change their ethnic essence. As evidence, Gil-White posed a series of questions about the ethnic identity of a child to his pastoralist informants in Mongolia. His evidence indicates that most Mongols believe that a Mongol baby (a child of two Mongols) adopted at birth by Kazakhs and raised as a Kazakh (and bearing all of the many Kazakh ethnic markers) will remain Mongol even if the child never learns of his parentage and spends his life thinking he is Kazakh. Using this essentialist thinking, individuals are able to effectively organize and extend information learned about individuals to whole groups of people. For example, if a Mongol were to learn that a particular Kazakh, or family of Kazakhs, are disgusted by the idea of eating yak head, this Mongol would be likely to infer that all Kazakhs are disgusted by eating yak head. Though such essentialist thinking is clearly not correct given contemporary scientific understandings, this kind of essence-based reasoning will produce a wide range of good inferences about human behavior. Interestingly, Gil-White's argument and evidence indicates that people think about (make inferences about) ethnic groups and biological species in remarkably similar ways. It seems possible that, in creating our ethnic psychology, natural selection made use of inferential processes developed for thinking and learning about biological species (Gil-White 2001).

Ethnicity and Cooperation

This book is about cooperation, not coordination, so how does ethnicity fit into our study of cooperation? The key to understanding this is realizing that the social norms theories we discussed in chapters 3 and 7 actually turn cooperative problems into coordination situations. Adding culturally transmittable punishing strategies, and thereby bringing norms into existence, often creates a situation in which the highest payoffs will go to those who avoid punishment—which usually involves obeying the norms. Different social groups will culturally evolve different norms, so individuals who want to have successful social interactions would do best to learn norms (and when to punish norm violators) from those most likely to have the right norms for the learner's group and to interact with those most likely to have the same norms. Otherwise, they will get gossiped about and punished. All this means that the ethnic preferences and psychology described above can apply equally well to both coordination situations and cooperative dilemmas involving social norms.

Moreover, the bias to interact with members of one's own ethnic group (to avoid miscoordination) means that the reputation and gossip networks, which influence both indirect reciprocity and the punishment of norm violators, won't generally extend very much outside of the ethnic group. Thus, the interactional bias created by an ethnic psychology will tend to limit the reach and effectiveness of avenues to cooperation. Groups like the Chaldeans rely on their tight social networks to pass reputational information and gossip that promotes cooperation and maintains norms through punishment and the threat of punishment. These avenues to cooperation are only as good, and as far-reaching, as the social networks that sustain them.

Ethnicity and Cooperation among the Chaldeans

Predictions

Our ethnographic work shows that Chaldeans tend to have positive feelings about their ethnic group that lead to a strong in-group preference and preferential interactions within their community. These interactions tend to be successful because they are well coordinated—people follow the same rules and norms in interactions. Through a feedback loop between warm feelings toward their own group and the success of the interactions owing to the high level of coordination, most interactions end up taking place within the ethnic group. In the remainder of this chapter, we argue that Chaldeans' strong ethnic identity has important effects on social behavior and costly cooperation in a manner consistent with the predictions laid out above and in chapter 3.

More precisely, we will address the following theoretical predictions here:

1. Chaldean ethnicity is marked principally by the costly or hard-to-fake cues of language and religion. In fact, the diffuse nature of life in Detroit has caused the Chaldean language to resurge and emerge as an important ethnic marker. Our theory predicts both the nature of the markers and its resurgence under the social conditions in Detroit.

2. Consistent with both theory and other empirical work on the psychology of ethnicity, Chaldeans tend to think that "being Chaldean" is partly transmitted down bloodlines. Although the Chaldeans do not entirely essentialize ethnicity, their beliefs are largely consistent with Gil-White's notion that humans tend to associate ethnicity with an essence (that causes outward behavior), which is passed down bloodlines.

3. Ethnic identity leads individuals to preferentially interact with other Chaldeans—that is, to interact with those showing the markers and avoid people without the markers. Both interviews and observations show a clear preference for interacting with coethnics in social and professional relationships.

4. In addition to this preferential interaction, we show that Chaldeans preferentially direct benefits toward other Chaldeans. We also demonstrate that not only do Chaldeans cooperate with coethnics but also that this cooperation is closely linked with their ethnic identity. Here, we present three indices measuring strengths of cooperation and ethnic identity and show that Chaldean cooperation is correlated with ethnic identity

and that people have a preference for cooperating with coethnics rather than general inclination for cooperation.

We also present evidence suggesting that one way in which Chaldeans maintain ethnic cooperation is by building the pursuit of self-interest into the social norm. Cooperation and social interaction should (from the perspective of most Chaldeans) be kept within the ethnic community but, within this context, Chaldeans can (and should) compete fiercely for business, good marriage partners, wealth, and prestige. This minimizes the conflict between individual and group interests and is likely a hallmark of effective human institution.

Background

Many members of the Chaldean community in metro Detroit were born in the United States or have lived there from a very young age. Over the years, the Chaldean community as a whole has become highly acculturated in many ways: nearly everyone speaks English (and many with no hint of an accent), most Chaldeans have moved out of the ethnically isolated Chaldean Town of Detroit and into middle- and upper-class suburbs, an increasing number of people attend university, and participation in the ethnic occupation of grocery store ownership and management is declining as Chaldeans pursue careers in a wide array of professions. With the exception of the elderly and people who do not work outside of the home, Chaldeans interact with non-Chaldeans on a daily basis— at school, at work, while shopping, and so on. At first glance, Chaldeans appear to be fully integrated into mainstream American society rather than a distinct community. However, despite their ability to function with complete competence in the broader society, Chaldeans perceive themselves as a distinct group and seek a degree of isolation in their social, professional, and religious lives.

The Psychology of Ethnicity among the Chaldeans

Chaldeans tend to see their ethnic identity as arising from a mix of descent (genes), language, and culture. From the perspective of Chaldeans, a person can be a Chaldean merely by being born to a Chaldean, but to be part of the community, one also needs to hold certain cultural beliefs and practices. A distinction is often made between people who are Chaldean and people who are Chaldean and part of the Chaldean community. In the minds of many Chaldeans, apparently there are people who are ethnically Chaldean (carry a Chaldean essence) and others who are also culturally Chaldean. A person can be an ethnic Chaldean without being a cultural Chaldean, but all cultural Chaldeans are also ethnic Chaldeans (with the possible exception of non-Chaldean babies adopted by Chaldeans, which we'll discuss below). Though not everyone agrees on which characteristics and behaviors are essential for full acceptance as a member of the group, most people agree on many of the traits, none of which is individually sufficient for membership.

Ethnographic work indicates that the most common characteristics that people see as important for being Chaldean include (1) belonging to the Chaldean Church; (2) speaking the Chaldean language; (3) having ancestors from Iraq (but not being Arab); and, more generally, (4) having (or conforming to) the morals, values, and traditions of Chaldeans. From our theoretical perspective, items 1, 2, and 3 are the markers, and 4 represents the underlying norms. In addition to these acquired characteristics, having been born to a Chaldean is also necessary. For example, a person not descended from a Chaldean who learns to speak Chaldean, marries a Chaldean, joins the Chaldean Church, cooks Chaldean food and associates exclusively with Chaldeans will not be considered a *real* Chaldean by most members of the community. Some Chaldeans designate these people (and there are some) as "honorary Chaldeans." But they remain distinct from true Chaldeans. Karen, a 28-year-old student who has lived in the United States since age 5, talked mockingly about her cousin's "American" wife. The wife attends all the family and community events and activities and dresses like a Chaldean, but the cousins make fun of her and call her "a Chaldean wannabe." Karen said, "She tries too hard, and it shows." However, the children of the "American wife" are considered Chaldean because their father is Chaldean (thus satisfying the descent requirement), and "because they live with their grandparents and are part of the Chaldean community.... The grandparents taught them the language, and they socialize with [their] Chaldean cousins" (thus satisfying the cultural requirements).

Religion

Belonging to the Chaldean Church is a salient trait of being Chaldean. For some, this is a crucial element of being Chaldean, whereas for others it is merely important. According to 25 percent of the people interviewed about this ($n = 36$; random sample), leaving the Chaldean Church disqualifies a person as Chaldean; they are no longer Chaldean in any sense of the word. A number of people Natalie spoke with had relatives who had become Jehovah's Witnesses. A relative of one of these converts was very emotional about the conversion and became visibly upset as she talked about it, saying "She [the convert] is not Chaldean as long as she doesn't come to our church." When asked whether a Chaldean who leaves the Chaldean Church is still Chaldean, a middle-aged grocer who moved to the United States at age 21 responded simply and emphatically, "Hell, no."

Although three-quarters of the community still perceive someone who converts out of the Chaldean religion as "Chaldean," nearly all of these people agree that a Chaldean who leaves the Chaldean Church ceases to be accepted in the community. Anne, in her 50s, came to the United States from Baghdad 25 years ago. Her reaction to converts reflects some of the exclusion they face: a convert is "not accepted by the community, we won't pray with him; but, we can't take away his Chaldean-ness." Her daughter, Sarah, a 21-year-old student born in the United States, emphasizes that a convert is still technically

Chaldean, but when talking about him people will clarify that he does not belong to the Chaldean Church, and he will not be part of the community. A 33-year-old social worker said that although he does not personally agree with social shunning, he believes that many people in the community would ostracize the convert and talk about what a crazy thing she had done. People who leave the church were described as "crazy" or "acting crazy" repeatedly during these extensive interviews. Sue, a professional woman married to a non-Chaldean, told Natalie that her cousin had become a Jehovah's Witness and that "people think he's nuts. . . . [They] see someone as mentally troubled if he leaves the Church." And a girl in her early 20s, who admits that she attends church only on major holidays, said that a man who left the church would still be Chaldean, "but I wouldn't date him, I would consider him a messed-up Chaldean." Thus converts become isolated from the community. It's as if converts have somehow betrayed their Chaldean essence, and they are punished in a manner worse than non-Chaldeans (non-Chaldeans aren't seen as deranged if they adopt another religion). Psychologically, Chaldeans who drop out of the church may be seen as "fakers" who carry many of the markers (language, dress) but not the norms.

The status of converts allows us to study the dichotomy between being Chaldean and being part of the Chaldean community. The idea of a Chaldean essence emerges as people talk about the enduring quality of being Chaldean even when the cultural practices associated with being Chaldean are absent. Retaining membership in the ethnic group when group norms are abandoned reflects a conceptualization of ethnicity as essentialized and a view that being Chaldean involves being born with an immutable Chaldean-ness. But ethnicity and culture are not the same thing, which is why a person who converts can retain his ethnicity but be excluded from the cultural group (i.e., he is Chaldean, but he is not part of our community).

The exclusion of a convert from the community is often coupled with strong, negative feelings toward the individual. Karen, the woman who referred to her cousin's wife as a "Chaldean wannabe," explained that someone who leaves the church is "still Chaldean, but the community looks down on the person. The culture shuns people who switch or leave the church." Or, put more emphatically by Rose, who herself has been a victim of social ostracism because of her divorce: "Everyone will point a finger at the person. He disgraces our beliefs and our values."

However, being a member of the Chaldean Church is not sufficient to make someone a Chaldean. Eighty percent of our informants said that a person of non-Chaldean descent who joins the Chaldean church (as well as acquiring other Chaldean traits like language and cooking Chaldean foods) is not a Chaldean, although the community may welcome him.

These patterns are consistent with the general theory developed earlier. Religion is a costly signal involving church attendance, monetary donations, and rituals. Leaving the religion (losing the ethnic marker) prompts people to infer that the person is crazy and lacks their values and norms. Thinking someone is crazy and lacking your values reduces their desirability as a business partner or mate. Because people don't see non-Chaldean religious converts as "crazy," there is something about being born Chaldean that makes one crazy to change religions. Consistent with our theory, losing the ethnic marker of religion doesn't (at least for three-quarters) strip one of his Chaldean-ness, which is treated as a kind of essence one gets from one's parents.

Historically, it is not particularly surprising that a group such as the Chaldeans have evolved a norm that punishes departures from the church. Remember, this cultural group has survived as a cultural unit for centuries in a part of the world dominated by Islam. Such a group would not have survived as an identifiable religious group if they had not developed norms that fortified the group against invading Islamic ideas.

Language

Many Chaldeans emphasize the ability to speak Chaldean as an important part of their culture. Their language is a source of immeasurable pride, and nearly every Chaldean that Natalie spoke with informed her that Chaldean was the language spoken by Jesus (Aramaic). Believing that the preservation of their language is essential for maintaining their culture, Chaldean language classes are attended by both children and adults. In Iraq, schools are taught in Arabic, so Chaldeans learned to speak their language but not to read or write it. Consequently, only a small minority of the community is literate in Chaldean, and the classes provide the opportunity to acquire this literacy.[3] As with many ethnic groups, a strong link exists between language and ethnic identity. Among the Chaldeans, speaking the language is a sort of litmus test for Chaldean-ness: Who is Chaldean? The people who speak Chaldean. For one grocer, whether or not someone speaks Chaldean influences whether or not he will help them. "I prefer to help people who speak my language," he said. We believe that the reason that he has this preference is not because of the language itself (he is fluent in English) but rather that speaking Chaldean acts as a signal of ethnic identity, and he prefers to help people whom he identifies as being the same as him.

As mentioned, language is a difficult ethnic cue to fake because it is extremely hard for anyone other than a native speaker to master the grammar, vocabulary, colloquialisms, accent and intonations of a language. The tie to the Chaldean language may partially result from its links to the Chaldean heritage and ancient origins, but also because it serves as a reliable cue of ethnicity, and thus, group membership. A woman in her forties who immigrated to the United States from Telkaif made it clear that she does not think well of people who, in her mind, try to fake the language cue and gain membership in the group under false pretenses. This woman is a teacher in a public school in an area with many Chaldean students. Some of the other teachers that come from villages in Iraq that are very near to Telkaif speak a dialect of Chaldean. Apparently, this woman does not consider these dialects to be the real Chaldean language and hence, to her, the speakers of these dialects are not Chaldean. With vehemence she proclaimed: "People call them Chaldean, but they are from the gutter.

I get angry and say to people, why do you call them Chaldean? They're not Chaldean, and they give a bad name to us." Although this reaction was unusually strong, it reflects the importance of maintaining the Chaldean language as a credible signal of ethnicity—and the fact that dialectical variation is sufficient to evoke strong emotions in our ethnic psychology.

The Chaldean language is used to create a feeling of solidarity, and it can be used to exclude outsiders—non-Chaldean speakers. Karen explained that all the Chaldeans in her group of friends speak English together but periodically switch to Chaldean when they tell jokes. Natalie often found herself in situations where everyone was speaking English, and then someone told a joke in Chaldean, and everyone burst out laughing. Whereas Natalie had felt part of the group one moment, she became a self-conscious outsider in an instant. Smiling, trying to pretend that she was enjoying the joke even though she had no idea what had been said, she had a heightened awareness that those who share the joke also share an identity and solidarity, whereas those who miss the joke are always on the outside. Language is a powerful tool that can simultaneously unite and divide.

The importance of speaking Chaldean was evident in the feelings of disgrace expressed by those who do not speak it. Sue, after saying that her children do not know Chaldean, added, "I am very ashamed of that." Sam brought his family to Detroit three years ago after living in Greece for 9 years. He says that the community criticizes him for speaking to his children in Arabic rather than Chaldean, explaining that choosing to teach your children a language other than Chaldean may be perceived as a rejection of the culture. However, Sam defends his choice, arguing, "There are 22 Arabic countries, and Chaldean is only spoken by one community within one country [Iraq]." Sam had told Natalie on an earlier occasion that a person who does not speak Chaldean cannot be Chaldean and that his children identify themselves as Greeks with Chaldean parents.

Speaking Chaldean may be such an important part of cultural identity because of its effectiveness as a marker.[4] As Sam pointed out, very few people speak Chaldean. This makes it a good trait for determining the group boundary, as well as for fostering a shared identity. Moreover, it is a difficult signal to fake because it takes a lot of time and effort for a person to learn a language that he didn't speak growing up. This is especially true for Chaldean because it is primarily an oral language, and it would be very hard to acquire without being immersed in a Chaldean community. Further, people can detect when someone isn't speaking Chaldean the way a "real" Chaldean ought to—as we saw with the teacher's reaction to the villagers who speak a Chaldean dialect. Preserving and promoting the Chaldean language serves as a reliable, hard-to-fake cue of group membership, which correlates with other underlying values and norms.

Without a proper theory, it may seem strange that the Chaldean language continues to develop such potency as a fundamental component of culture, heritage, and identity, given that most recent immigrants don't speak a word of it when they arrive in the United States. Surely the Arabic-speaking Chaldeans who have been residing in Iraq are no less Chaldean than the American-born, Chaldean-speaking Chaldeans. Recent immigrants usually come from Baghdad or Mosul, large Iraqi cities where they were fully immersed in Arabic. In contrast, the earlier immigrants came from northern villages where Chaldean was spoken in all contexts other than school. The descendants of the early immigrants learned Chaldean from their parents, whereas the Chaldeans who remained in Iraq moved to the cities and lost their language over time. Because Chaldean became the lingua franca for the Detroit community, many of the new immigrants find it necessary to learn Chaldean to communicate with earlier immigrants and their descendants. A young doctor who came to Southfield from Baghdad in 1992 has started to acquire Chaldean from his patients, many of whom came from Telkaif. A couple from Baghdad who came to the United States in 1969 arrived not speaking any Chaldean but learned it here so that they could communicate with the other Chaldeans. Though the Chaldean language remains strongly associated with cultural identity, many people attend mass in Arabic so that they can read the prayers. It is ironic that although religion and language are two of the most important characteristics in defining the Chaldean identity, many Chaldeans say their prayers in Arabic.

Our ethnicity theory may help explain why the Chaldean language has reasserted itself. Modeling work shows that in situations involving a high level of intermixing between groups with different social norms and little spatial structure, markers need to be both readily identifiable and highly reliable cues of underlying norms. In situations in which people may be spatially dispersed into distinctive neighborhoods, having a multitude of semi-reliable markers can be sufficient for determining group membership and ethnic identity. In contrast, a metropolitan, multiethnic situation places selective cultural learning pressure on a few "hard=to-fake" cues, such as language. Keep in mind that Detroit's Arab community is much larger than the Chaldean community, and Chaldeans would want to be able to quickly differentiate one of their own from the Arabs. Of course, some Chaldeans who could have been good people to interact with may be avoided because of the emphasis on this cue, but such is the price of reliable cues.

Chaldeans as Non-Arabs

In Iraq, where most of the country is Arabic, Chaldeans constitute a small minority of the population. In metro Detroit, Chaldeans are again a minority compared to the Arabs, who have a population of more than 300,000. Perhaps because of the long history of living in the shadow of larger Arabic communities, Chaldeans define themselves in contrast to Arabs. To many in the community, being a Chaldean means not being an Arab. This may serve to unify the community in two ways. First, it creates a shared trait, that of non-Arab. Second, it strengthens the in-group bond by creating a well-defined out-group, the Arabs. The Arabs are often seen as the enemy because they conquered the Chaldeans in ancient times and are the dominant population in modern-day Iraq, oppressing and discriminating against the Chaldeans. Most people explain that Chaldeans are not Arabs and note that the two groups speak different languages, practice different religions, and have different cultures and histories. They also quickly point out that the Chaldeans predate the Arabs. Two-thirds of Chaldeans say that they are not Arabs, and many of these people say that it is offensive to call a Chaldean an Arab. The remaining third of the community does identify itself as Arab, and these people say that they all come from the Middle East and speak Arabic. Remember, the Chaldeans learned to read and write Arabic at school in Iraq, and some who grew up in the large cities only speak Arabic. Both camps, those who identify as Arabs and those who do not, feel passionately about this aspect of their identity, and some Chaldeans seemed angry that Natalie would ask them whether Chaldeans are Arabs because they saw the answer as obvious and seemed to think that she was questioning their identity. People on each side of the issue fall victim to the false consensus effect (Ross et al. 1977), insisting that everyone in the community feels the same way they do.

Ethnic and Conformist Learning Biases Lead to a Range of Weak Ethnic Markers

Though a range of empirical evidence indicates that people use both conformist transmission and an ethnic bias in figuring out whom to learn from, we don't know exactly how the learning biases interact. One possibility is that people use an ethnic bias (and other self-similarity biases, such as sex) to select among individuals to pay attention to for cultural learning. Then, in figuring out what to adopt from these people, individuals use conformist and prestige-biased forms of cultural learning. When differences in prestige do not correlate with differences in behavior/beliefs, individuals tend to adopt behaviors held by the majority of their ethnic group (and also consistent with their sex and age cohorts). A variety of combinations, including the ones mentioned above, can generate broader patterns of similarity among coethnics, which may act as weak ethnic markers.

Shared behaviors and practices abound in the Chaldean community. Although many are not linked to traditional Chaldean culture, they serve as weak ethnic markers that foster group identity and help to define group membership. For example, Chaldeans frequent the same restaurants, coffee shops and bars, prepare Chaldean foods and drinks in a similar way, have a distinct style of interaction, and have a common way of hosting or welcoming visitors. A woman illustrated the homogeneity of Chaldean life when she described the parks that Chaldeans tend to visit on the weekends. "All Chaldeans used to go to Kensington Park, then they switched to Stoney Creek," she said. "Now people go to both." Cass Lake is also very popular, and "95 percent of the people there on Sundays are Chaldean."

Many other shared behaviors, cutting across all areas of life, unite the group. One highly visible domain of conformity is dress and attire, which has become a vehicle for creating a sense of group cohesion. As a non-Chaldean intimately familiar with the community put it, "the Chaldeans are always dressed to the nines." The use of fashion as a homogenizing force is greatest among the younger generations (children, teens, and adults in their twenties and thirties). However, it is not just that people are fashionable. Young Chaldeans are all fashionable in exactly the same way. If Chaldeans are concerned with individual variation in attire, the differences could only be described as subtle. Over the course of a year, Natalie attended a weekly meeting of men and women in their twenties, and each week she noted what people were wearing (especially the women) because she was keenly aware of how differently she was dressed from everybody else. Most weeks, Natalie came in jeans, and with the occasional exception of one woman, she was the only female wearing them. This was not a result of coming underdressed to a more formal event—the meetings were very casual, with many of the group members coming to the meeting after school. In addition to extreme uniformity in footwear (black leather shoes or boots with high, thick heels) and shirts (usually tight), the married and engaged women wore nearly identical engagement rings. Natalie, being herself engaged at the time, was very aware of engagement rings and had gotten in the habit of checking out rings worn by friends and strangers. She learned that there is plenty of variation in ring styles, so she was all the more surprised that the women at the meeting displayed none of this variation. Intrigued, Natalie asked a few people how they chose their rings and whether the man picked it alone or with the woman's input. One man from the group explained that the couple often looks at rings together prior to the engagement so the woman can tell the man which ring she likes, and the "girls are hip, they want the same things [as their friends], and they know what their [Chaldean] friends got." Through a strong conformist effect, the women end up dressing alike, right down to the accessories. The same is true for the men, with most wearing dark pants, black shoes, and similar stylish shirts. In a sense, the Chaldeans have created a dress code, or uniform that may help to create a feeling of unity and that marks them as members of the same group. Unlike language, the dress code can be copied by non-Chaldeans and is a much less reliable marker of group membership.

Chaldean by Descent

Above, we discussed the dual requirement for identification as a Chaldean: sharing certain cultural characteristics and being born to a Chaldean. However, certain rules of descent must be met in order to satisfy the inheritance component.[5] To explore this, Chaldean informants ($n = 36$) were given three scenarios regarding ethnic identity. In the first scenario, informants were asked, "If a child has a Chaldean father and a non-Chaldean mother, what is the child?" Next, "If a child had a non-Chaldean father and a Chaldean mother, what is the child?" And, finally, "If a child of non-Chaldean parents is adopted by Chaldean parents and acquires Chaldean language, customs, and religion, what is the child?" The answers were free-response, but all were immediately codable into three categories. The results are shown in figure 9.1. Consistent with Chaldean beliefs about patrilineal inheritance, more than 80 percent of our informants maintain that if the child had a Chaldean father, he or she was Chaldean (we did not specify a sex for the

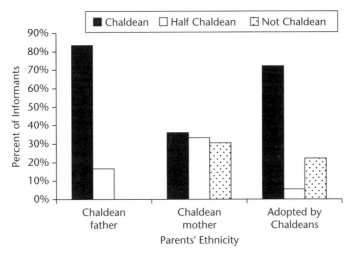

Figure 9.1 Child's ethnicity when only the father is Chaldean, only the mother is Chaldean, and when a non-Chaldean baby is adopted by Chaldeans ($n = 36$).

child in the question). In contrast, people were quite split when the mother was Chaldean, but the father was not. Answers in this case showed a nearly equal division between "half Chaldean," "Chaldean," and "not Chaldean." This work suggests that Chaldeans, like other groups, bring an essentialist bias to understandings of ethnicity, but that the application of this bias is strongly influenced by culturally transmitted rules of descent (patrilineal inheritance in this case).

Throughout our study of ethnicity among the Chaldeans, people agree that a person cannot be a "real" Chaldean unless he was born to a Chaldean (or at least to a Chaldean man), regardless of the acquisition of cultural traits. Given this robust finding, the answer to our third question (from above) seems anomalous.[6] Seventy-two percent of people interviewed said that a non-Chaldean baby adopted by a Chaldean family would be Chaldean if it was raised within the Chaldean community. However, many of the answers were qualified. Physical appearance influences whether or not an adopted child will be considered Chaldean and/or accepted in the community. One man said that people would recognize the child as Chaldean but would not accept her as part of the community if she looked different. Sarah stated that she would not consider the child to be Chaldean and would welcome her in the group only if she looked Chaldean. Her cousin adopted a black child, she said, and she does not consider him Chaldean. Interestingly, Sarah does not get this attitude from her mother, Anne, who does accept adopted children as Chaldean. On many topics related to being Chaldean, Sarah takes a more purist view than her mother. Sue said she would neither consider an adopted child as Chaldean nor accept her in the community because she would always know by looking at her that she wasn't Chaldean. She claims that she can identify Chaldeans by appearance, even distinguishing between Arabs and Chaldeans. We are skeptical as to the accuracy of her identifications because other people who told Natalie that they can identify Chaldeans on sight also thought that Natalie was Chaldean (Natalie is of Eastern European Jewish descent). Sue is married to a non-Chaldean, and she considers her own children to be "half Chaldean." For some people, an adopted child will not be considered Chaldean regardless of what he looks like. Jennifer is an American-born 31-year-old with a university degree who works in the entertainment industry. She said that she would not recognize an adopted child as Chaldean regardless of what he looks like, explaining that people would always know where the child came from and the community "would put a stigma on him and always identify him as not one of us."

Given that these are fairly wealthy, educated people, we think the interesting part of the results from question #3 is that nearly 30 percent of Chaldeans would not see an adopted child who acquired the language, customs, and the religion as a Chaldean. By the same token, without knowing any details about the child's upbringing or linguistic talents, more than 80 percent of informants readily assigned a child born to a Chaldean man as "Chaldean." Some kind of essentializing psychological tendency (whether the product of genetic or cultural transmission) is required to explain our pattern of results.

Overall, our results show that, for Chaldeans, determining one's ethnic identity is a complex matter, and substantial variation in the details of ethnic identification exist within the Chaldean community. In general, this process involves integrating descent and morphological qualities with such cultural elements as language and religion. Even the heuristic that 'a real Chaldean has Chaldean parents' has variability and exceptions. While most people agree that a person with a Chaldean father is a Chaldean, there is less consensus about a person's ethnicity when only his mother is Chaldean; moreover, the descent rule appears to be tossed out altogether in cases of adoption, but morphological similarities enter the assessment in these cases (note that morphology is apparently never an issue when descent is known).

In-Group Preference: Social Domains

Chaldeans show a strong preference to form relationships with other members of their ethnic group, both socially and professionally. Though Chaldeans interact with non-Chaldeans daily, they actively seek out other members of their community for socializing. When asked about the ethnicity of their friends, 67 percent of respondents stated that all of their friends are Chaldean (fig. 9.2).

To get more exact information on the extent to which Chaldeans preferentially socialize with coethnics, Natalie asked people to recall the last three times they got together with friends, whether it was for dinner, coffee, visiting in the home, going to the movies, and so on, and to tell her how many of those events were exclusively with other Chaldeans. The percentage of all-Chaldean socializing in all three of their last social engagements (64 percent) supports the finding that most Chaldeans only have coethnic friends and that most socializing takes place among Chaldeans (fig. 9.3). A few people who reported having only

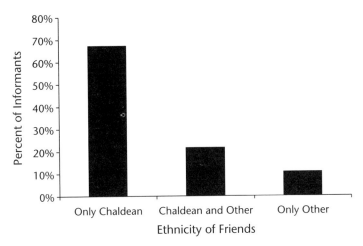

Figure 9.2 Ethnicity of informants' friends ($n = 49$).

Chaldean friends socialized with non-Chaldeans at work-related social events, but they consider these people to be coworkers rather than friends.

Although the number of friendships with non-Chaldeans is low, the quality and emotional importance of these relationships can be very high. Several people responded that most of their friends are Chaldean, but that their closest friends are non-Chaldean. Mark, a 29-year-old banker, said that 95 percent of his friends are Chaldean but that his best friend is not, and two women (a 29-year-old psychologist who migrated from Iraq at age 11 and a 36-year-old grocer who came to the United States at age 3) with primarily Chaldean friends reported that their best friends are Jewish and Italian, respectively. When people have non-Chaldean friends, they tend to be "ethnic". Although a few people

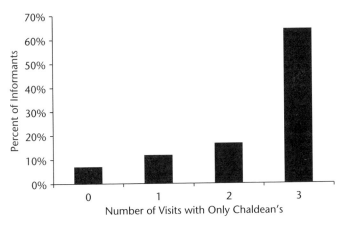

Figure 9.3 The number of the last three socializing events that were exclusively with Chaldeans ($n = 42$).

have what they call "American" friends—white Christians of European descent—most non-Chaldean friends descend from Middle Eastern groups, such as Assyrian[7] and Arabic or, to a lesser extent, Mediterranean groups, such as Italian and Greek. Chaldeans report good relationships with non-Chaldean coworkers, but they generally do not socialize with them outside of work. Michael is a 50-year-old financial advisor who has lived in metro Detroit since age 25. He works easily with non-Chaldeans and enjoys these professional interactions, "but when it's time to go out at night, it's with Chaldeans." Michael does not feel comfortable in social situations with his colleagues, explaining that "what interests me doesn't interest them, what interests them doesn't interest me, so how could I be part of their country club?" We suspect that the pattern of having one's closest friends (when they are non-Chaldean) be from a Mediterranean immigrant population reflects two opposing forces: (1) the tremendous reputation-related and conformity pressures people feel around other Chaldeans (that makes it difficult for some to "be themselves"), and (2) the tendency to seek out culturally similar friends (Assyrians and Arabs, for example, share many cultural preferences with Chaldeans, like foods) who aren't part of the dense Chaldean gossip network.

Marriages are even more exclusively Chaldean than friendships. Of the people Natalie interviewed who are currently married or engaged ($n = 26$), 89 percent were married/engaged to a Chaldean, with no differences in rates of endogenous marriage by occupation. Sengstock reports a nearly identical rate of coethnic marriage based on her 1963 survey of Detroit's Chaldean community, at which time she found that 88 percent of Chaldeans employed in groceries and related businesses married endogenously, as did 70 percent of non-grocers (1999: 54). Both results confirm the extreme rate of endogenous marriage that endures even in the face of many years of acculturation.

It is important to note that there is a bias in the high rate of in-group marriage among our informants. As we said earlier, Natalie conducted her interviews with a random sample selected from the 1998 Chaldean Directory, a subset of nonrandomly selected individuals whom she met through various Chaldean organizations, and a small number of individuals she met through other people whom she had interviewed. This means that, with the exception of the people Natalie met through other informants, all of the people interviewed identify themselves, to varying degrees, as part of the Chaldean community.[8] Since Chaldeans who marry non-Chaldeans are more likely to be distanced from the community than are Chaldeans who marry within the ethnic group, our sample does not accurately reflect the number of ethnic Chaldeans who married non-Chaldeans and are now outside (or on the margins) of the community. The higher overall rate of out-group marriage became evident when informants were asked if they have relatives who married non-Chaldeans. Of 46 informants, 54 percent have at least one relative who married exogamously.

The Chaldean preference for coethnic interaction can also be seen in the patterns of residency location, which reflects a desire to stay near family. Many people have at least one member of their immediate family living within ten minutes of them, and a number of Chaldeans live in the same block as their relatives. Both young and old seek to live close to their family. For example, back in the 1970s, a grocer, now age 70, and his four brothers-in-law bought adjacent lots and built homes next door to each other. The siblings and their families see each other daily. Recently, a 28-year-old professional bought a house next door to his brother. These two cases are hardly the exception, with numerous Chaldeans living next door to (or at least on the same street as) parents, siblings, grandparents, and cousins. A middle-aged teacher who migrated from Iraq in her twenties lives next door to her mother, aunt, sister, and nieces and nephews (all of whom live together), and two doors away from her cousins. Overall, nearly everyone has some relatives who live less than ten minutes away, with the most distant (in-state) relatives living approximately thirty minutes away. The clustering of Chaldean homes reflects the general preference for interaction among Chaldeans (especially with kin), and further increases the frequency of socializing within the ethnic community.

In-Group Preference in Professional Domains

When given the opportunity, Chaldeans seek to participate in business and professional relationships with coethnics. These in-group relationships include clients hiring Chaldean accountants and lawyers, patients going to Chaldean doctors, grocers buying from Chaldean distributors (and, conversely, distributors selling to Chaldean grocers), and Chaldeans opening businesses together or working at the same companies (although this is usually with relatives, rather than coethnics in general). In many cases, the preference to form endogenous relationships also leads to cooperation and creates individual and group benefits. As you'll see, it becomes tricky to parse the in-group preference from the resultant cooperation because they are two sides of the same coin.

Although the number of Chaldean professionals (doctors, lawyers, accountants, dentists, etc.) is rapidly increasing, they still represent a very small fraction of the total number of professionals in metro Detroit. Looking solely at the available professionals in the area, we would expect that most Chaldeans would hire the services of non-Chaldeans because there are so many more of them (Chaldeans make up approximately 4.8 percent of the metro Detroit population).[9] However, this is not the case. Depending on the service, between 42 percent and 63 percent of Chaldeans use the services of coethnic professionals (fig. 9.4).

Chaldeans usually give one of three reasons to explain why they prefer the services of another member of their ethnic group. First, some people say they want to help other Chaldeans, and they can do this by giving their business to them and thereby keeping money in the community. Lida is an unmarried, 44-year-old grocer who has lived in the United States for 30 years. Her dentist, lawyer, and accountant are all Chaldean, although she sees a non-Chaldean doctor. She says that she uses coethnic professionals in order to support members of her ethnic group. "I see other communities, especially the Jewish community, tend to patronize their own," she said. "I see how the Jewish community has prospered because they patron other Jews. So why should I give money to others? It's better to keep it within our own community."

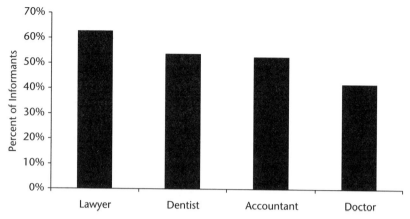

Figure 9.4 Percentage of Chaldean informants who use a coethnic lawyer ($n = 27$), dentist ($n = 26$), accountant ($n = 19$), or doctor ($n = 31$). Analysis includes only informants who use these services.

Second, many Chaldeans say they feel more comfortable interacting with other Chaldeans because of the shared language, culture, and interactional norms. In many of these cases, the client/patient is fluent in English but prefers to speak in Chaldean or at least have that option. In addition to the increased comfort level, empathy is cited as a factor; people say that a coethnic profession-al understands their problems better than a non-Chaldean. Many grocers use a Chaldean lawyer for legal issues related to business, and some say this is because Chaldean lawyers understand the grocery business better than non-Chaldeans. People also favor Chaldean lawyers for personal matters, again saying that a person with the same culture will better understand their problems. Jane, a 28-year-old American-born computer programmer, explained that "Chaldean law-yers can be more helpful because they are familiar with how we are, what we do. They know us better." The preference for Chaldean lawyers is also evident by looking at Chaldean lawyers' clients. While 63% of Chaldeans use a co-ethnic lawyer, some Chaldean lawyers report that 80–90 percent of their clients are Chaldean. Other lawyers have fewer Chaldean clients (10–50 percent), but some of these say they relied more heavily on Chaldean clients when they were starting out. One lawyer who now has only 10 percent Chaldean clients said that 90 percent of his clients were Chaldean when he started his practice. He explained that a loyal clientele early on can be the key to a firm's future success.

Third, people often choose a service provider based on recommendations from friends and relatives, or by choosing among people they know. Since most people have Chaldean friends, following a recommendation or picking among friends will often lead to the hiring of coethnics. Michael, a financial advisor, said that in choosing professionals for his own use, he goes to Chaldeans because he knows who is good. "I know the Chaldeans," he said, "so I know which ones are smart and qualified. If I go to an 'American,' then I don't know who is good." Similarly, Amy, a 30-year-old engineer who immigrated five years ago said: "You hear from people who is good. If I go to an 'American', I wouldn't know if he was good or not." Potential benefits from going to a professional who is also a friend, or who moves in the same social circles, include getting better service. According to Sue, she—and others—prefer to go to a Chaldean doctor "because the patients and doctors know each other, so you get better treatment and bedside manners." For all of these reasons, many Chaldeans explain that they favor using coethnic professionals.

Although more Chaldeans use coethnic professional services than would be expected given the number of Chaldean professionals, a nontrivial share of the community does not seek out the services of coethnics. Some of these people have neutral preferences—they do not care if the service provider is Chaldean or non-Chaldean, as long as he does his job well (the "merit model," which likely comes from cultural contact with the American majority culture).

However, there is also a small segment of the community that actively avoids Chaldean professionals. There are two distinct reasons that people give for preferring the services of non-Chaldeans: (1) qualifications, and (2) confidentiality. Two women expressed concern with the qualification of Chaldean doctors trained in Iraq, because they believe that these doctors are not as well trained as American doctors and that they are less familiar with the latest technology and medical advances. Notably, these women do prefer to use Chaldean professionals in other areas.

A broader concern is that of trust and confidentiality. Because of the close-knit nature of the community and the abundance of gossip, people fear that their lawyer, doctor, psychiatrist, financial advisor, and so on will tell their friends and family about things that were discussed in confidence. Although professionals are just that, and we do not believe that they break the confidences of their patients/clients, it is easy to understand why people have this concern. In a community in which most people know each other and gossip spreads like wildfire, it becomes difficult to believe that the usual cultural norms do not apply in their professional relationships. Receiving services from someone familiar increases the fear that personal information will be leaked to other Chaldeans. For example, a psychologist said that it is common for him to attend parties with his patients so he interacts with his patients in two roles—as a professional confidant and an acquaintance/friend. For some Chaldeans, the concern with trust and confidentiality is sufficiently great that they avoid the services of Chaldean professionals. Nevertheless, despite such concerns, most Chaldeans still prefer to obtain professional services from other Chaldeans.

In-Group Preference in Business

Chaldean grocers and wholesalers prefer to do business with coethnics. Since there are so many more Chaldean grocers than wholesalers, the degree to which each relies on the other is unbalanced, although both appear to be partial to these endogenous relationships. Chaldean wholesalers can have a clientele that consists almost entirely of Chaldean grocers, but no Chaldean grocer can stock his store by relying only on Chaldean wholesalers because the breadth of the stock requires small shipments from many distributors. However, when given the opportunity to buy products from a Chaldean wholesaler, most grocers jump at the chance. The reasons given are very similar to those given by clients and patients who prefer Chaldean professionals. The grocers say that the relationships with the coethnic wholesalers are friendlier and more comfortable: they both speak Chaldean, they know each other outside of business so they help each other more than they would a stranger, they understand each other in terms of the culture and the business, and they may even be relatives (often, cousins). Although all of these factors are important, every grocer who buys from a Chaldean wholesaler makes clear that he does so only if the Chaldean's prices are "competitive," or at least are close to competitive. As above, however, a minority of Chaldean grocers deploy the "merit model" and claim to choose wholesalers solely on their honesty, reliability, and prices, regardless of ethnicity.

Wholesalers, on the other hand, say that the fact that most of their clients are Chaldean does not reflect a preference for doing business with Chaldeans but is a result of Chaldeans' owning nearly all the grocery stores in Detroit. Though a grocer can choose whether to buy from a Chaldean or non-Chaldean wholesaler, a wholesaler working in Detroit has very little choice in whether his customers are Chaldean. The wholesalers with whom Natalie spoke have, on average, an 85 percent to 90 percent Chaldean client base. Although the wholesalers explain that selling to Chaldean grocers is "natural" given the number of stores they own, the more relaxed relationship with the coethnic grocers does seem to reveal a preference for these relationships, or at least, special treatment. For instance, a wholesaler described the difference between extending credit to a Chaldean versus a non-Chaldean grocer. He sells to Chaldeans whom he knows or whose family he knows, and thus he can select to do business with people who have good reputations. Consequently, he feels secure in giving them credit ("business is based on a handshake") and does not require any signed contract, collateral, or postdated checks—things he does require of non-Chaldeans. Too, the Chaldean wholesalers make the paperwork easier than do the non-Chaldean suppliers. "They just hand you a receipt with the price and what was sold," the grocer said. "Non-Chaldeans have printed receipts, invoices, and faxes. It's a much simpler system with the Chaldeans." Only a few wholesalers said that there is no difference in the way that they do business with Chaldeans and non-Chaldeans.

Although the Chaldean clients of financial investors may not be easy to work with and may expect a lot for having given their business to the investor, the investors need their Chaldean clients and are thankful for their business. Robert, a middle-aged, Iraqi-born investor, has become very successful and owes much of his success to the patronage of the Chaldean community. When he started out, he had to recruit all of his clients. To do this, he turned to his community and got friends and family to invest with him. As his reputation grew, other Chaldeans came to him, and now he no longer needs to actively recruit clients. He maintains a client base that is 50 percent Chaldean. Networking within his community has been a great source of business, as he explained that he "is highly visible in the community and gets business by word of mouth. At parties, people are always coming up to me and asking about starting accounts with me or asking about how their existing accounts are doing."

Individual Interests, Conflict, and Cooperation

Though many Chaldeans prefer interacting with other Chaldeans, there are clearly costs and conflicts associated with ethnically biased cooperation, and sometimes people "do worse" by favoring coethnics. The most common examples of "doing worse" come from relationships between lawyers and their clients. When taking business from Chaldean clients, Chaldean lawyers expect to spend larger amounts of time with the client (greatly exceeding the amount of time that they would devote to a similar case for a non-Chaldean), debate with the client about rates since such clients frequently expect a discount, and defer receiving payment until the case is complete—and even then it may be months before the payment is made. All this time, the lawyer struggles to protect the relationship and abide by the norms of the culture. This means that the lawyer can't call the client and tell him to pay because it would seem disrespectful. A lawyer with a large Chaldean client base says that asking to be paid is difficult to do without offending the client; rather than calling a client, she said, she sends a polite letter asking to be paid and then waits. Though all the lawyers seem grateful to their Chaldean clients for their role in helping the lawyer get started professionally, many also seem delighted to attract non-Chaldean clients once they become successful. As one lawyer who started his practice with nearly all Chaldean clients said, "Chaldean clients provide a great base, they are very loyal." But after becoming well established, this lawyer reduced his number of Chaldean clients because he found them too demanding.

Chaldean clients also challenge their coethnic financial investors/advisors. According to advisors, investing is new to Chaldeans, and they are less educated and sophisticated about investments than they are in other areas. These clients call frequently with questions and queries for updates, so that they require much more time than non-Chaldean clients. Robert said that his coethnic clients call three to four times more often than his other clients, and another investor recently ended a relationship with one of his Chaldean clients who, he said, was too demanding.

In the grocery business, both grocers and wholesalers try to get the best possible deals for themselves. Some wholesalers try to pressure grocers to buy more of their products, as one grocer explained, "Sometimes Chaldean vendors take advantage of you because you're Chaldean. They try to get special favors, ask you to take a few more things." However, the wholesalers also provide extra benefits to the grocers, such as making suggestions about what to sell and disclosing information about what is selling well at other stores. Although the wholesaler may try to push more products than the grocer needs, overall he strives to make the relationship beneficial for the grocer in order to keep his business. As another grocer put it, "They [suppliers] go the extra mile for us, and we oblige them any way we can." This summarizes the relationship nicely: the grocer tries to get the best deal and service he can, the wholesaler tries to sell as much as he can, and in the end (usually) both parties are satisfied and cooperate with each other in order to sustain the mutually beneficial relationship.

Summary

Chaldeans derive a strong identity from their ethnicity and culture. As a cause and an effect of group identity, Chaldeans have a greater comfort level with other members of their community that leads to a preference to form relationships, both personal and professional, with coethnics. People say that they prefer coethnic relationships because they are familiar with each other's customs, language, and business. Dealing with people who share these characteristics may lead to more successful interactions and a greater comfort level and overall satisfaction with the interactions. The success of these well-coordinated relationships may further increase the intragroup preference. People seek out others with their same cultural identity because of a generalized favoritism for the in-group, and the success of these interactions creates a feedback loop that augments the initial preference. Additionally, simply having a high level of in-group interactions further enhances the cultural group identity, which in turn strengthens the preference to function within the community. This interplay between interaction and identity creates a strong ethnic cohesion that has real consequences in the pattern of all types of relationships, including cooperation-based relationships.

Chaldeans Preferentially Cooperate with Coethnics

So far, we have shown that Chaldeans prefer to interact and engage in social and professional relationships with coethnics, and that this preference seems to be linked to a strong ethnic or cultural identity. However, Chaldeans not only preferentially interact with each other, they also tend to direct cooperative behaviors and benefits toward members of their ethnic group. Moreover, Chaldeans are able to avoid some of the conflicts between individual and group interests that occur in cooperative situations by relying on norms of interaction that enable individuals to competitively pursue their own self-interest while simultaneously benefiting the group as a whole. The pursuit of self-interested goals helps mitigate the desire to free-ride, and thus makes it easier to sustain the group-beneficial norms (i.e., the preferential relationships within the ethnic community). Since each individual in a relationship strives to get the best deal for himself, conflict and competition exists within the framework of a cooperative relationship. In short, Chaldean norms have evolved to manage self-interested motives, thereby reducing the tendency of self-interest to erode certain forms of pro-group prosociality.

When Chaldeans cooperate with nonrelatives, including strangers, they are most likely to do so with another Chaldean or as part of a Chaldean group.

If we look at involvement in a group, such as participating at the church, belonging to a professional organization, doing volunteer work through an organization, joining a social club, and so on, it becomes strikingly evident that Chaldeans join groups that have Chaldeans as their members. Chaldeans prefer groups that help other Chaldeans. Of everyone interviewed, 75 percent belong to at least one social group or organization. Eighty-five percent of these people belong to a Chaldean group, although of those who belong to a Chaldean group, 57 percent also report involvement in a non-Chaldean organization.

Politics

When we look at one particular domain of cooperation, political involvement, the same tendency to support and promote Chaldean interests emerges. Although Chaldeans claim that their community is not politically active, our sample revealed quite a high level of participation among adults of all ages. Nearly 32 percent of Chaldeans have protested at least one political issue, and have done so in ways that range from writing a letter to the local newspaper to traveling to Washington, D.C. to join a protest march. Of these protesters, 60 percent protested a Chaldean-related issue, with protests against U.S. sanctions on Iraq being the most popular issue.

Similarly, when Chaldeans work on a political campaign, the candidate is usually Chaldean. Among our informants, 58 percent worked on at least one political campaign, trying to get people elected to offices ranging from local trustee or district judge to senator. Approximately 40 percent of people who worked on a campaign helped in the 1994 senatorial campaign of Spencer Abraham, a Republican "Arab." Most of these same people joined the majority of the remaining 60 percent of campaigners in working to get Chaldeans elected to office. In the most recent campaign at the time of the research (November 2000), a Chaldean lawyer (Diane D'Agostini) was running for district judge. Many Chaldeans participated in this campaign, making phone calls, handing out fliers, and putting up signs on people's lawns.

Chaldeans participating in politics have consciously decided to help members of their ethnic group. People were very blunt about the need to support coethnic candidates (as noted earlier with interracial marriage, Chaldeans reveal none of the politically correct discourse found in and around universities). A representative from D'Agostini's campaign spoke to a group of Chaldeans and proclaimed that they must help in the campaign because they need to get Chaldeans into office, and it is necessary to support their own. A woman who has been extremely active in numerous campaigns told me that she has made a decision to stop working on campaigns for non-Chaldeans and focus her time and effort on getting Chaldeans elected into office. This pattern of involvement shows the same thing that we've seen throughout: when a Chaldean gets involved, whether it be in a personal dyadic relationship or political involvement, she is most likely to cooperate, help, and interact with other Chaldeans.

Preference for cooperation and interaction with coethnics benefits the group as a whole. The most obvious way that the community accrues benefits is through the accumulation of wealth, and consequently status and political influence. The increased wealth of individual group members helps the entire group as the affluent individuals give money back to the community (by spending money at Chaldean businesses/services, making loans, and donating to charity), and as they acquire power beyond the ethnic group. As we said earlier, Chaldeans are becoming increasingly politically savvy as they recognize the value of political donations and support, which are made possible by their growing wealth. Chaldeans also gain influence via their economic strength as owners of businesses and consumers of products and services. As the power of some members of the group rises, the benefits flow to all members of the community. For example, Chaldean influence led the city government to offer large financial incentives exclusively to Chaldeans to redevelop Chaldean Town.

Although increased wealth of individuals offers benefits to the group, the drive to acquire wealth and aid the community seems to be strongly fueled by a desire for personal status. The community highly values financial success such that respect and prestige accrues to the possessor of money and assets. Consequently, individuals strive to become rich, choosing occupations that provide large incomes, and make highly visible displays of their wealth. As individuals work to fulfill their personal goals of affluence, the community as a whole gains from the benefits that result from this wealth and influence. Thus, when Chaldean patients visit Chaldean doctors, or Chaldean clients hire Chaldean lawyers and financial investors, these decisions result in increases in the average wealth of the community by keeping money within the ethnic group. However, while there is a notion of helping other Chaldeans, there is no sense in which everyone should be equal in terms of success or wealth. Cooperation, via endogenous business and political relationships, helps individuals acquire prosperity and indirectly leads to group-level benefits.

Although the individual professionals and the group as a whole benefit from these interactions, there is also a self-interested incentive for the consumer to free-ride (i.e., to do business with a non-Chaldean; as a simple matter of statistics, a person whose goal is to get the best deal, hire the best professional, and so on will usually end up doing business with a non-Chaldean). Because of this conflict, preferential interaction with coethnics can create a cooperative dilemma. Costs that the consumer incurs may include doing business with someone who charges more than an outsider, who may be located less conveniently, or who may be less competent than an outsider. Given the relatively few Chaldean professionals compared to the total number of professionals in metro Detroit, it seems quite likely that a more affordable or more skilled non-Chaldean is overlooked by narrowing one's options solely to Chaldeans. This is seen by the number of grocers who say that they buy from Chaldean wholesalers even if they charge slightly more than non-Chaldean competitors (if the price difference becomes too great, then they switch to the non-Chaldean). By individual consumers accepting the costs of doing business within the ethnic community, the group prospers financially.

Ethnic Identity and Cooperation

Up to this point, we have described the Chaldean ethnic identity and the tendency for Chaldeans to preferentially interact, and cooperate, with members of their ethnic group. Now, we examine more closely the relationship between ethnic identity and cooperation by considering two key questions: (1) Does the strength of one's identification with the ethnic group lead to more cooperation with other members of the community? and (2) Is the tendency to cooperate a generalized disposition that effects all social domains, or is it targeted and context specific (as we argued in the last chapter)?

To explore this, we will look at a variety of real-life prosocial behaviors among both Chaldeans and non-Chaldeans and a combined measure of ethnic identification. Our discussion follows three steps. First, we show that ethnic identification (EI) predicts (correlates with) prosociality and preferential interaction with other Chaldeans, even when sex, age, and income are controlled for. Second, we demonstrate that cooperation with Chaldeans (CC) and non-Chaldeans (CNC) is uncorrelated, and that EI does not predict CNC. Thus, we are not measuring a general tendency to cooperate, but an ethnic psychology that influences social behavior. Finally, we'll explore the factors that may lead to strong ethnic identification.

The Indices

To investigate the relationship between ethnic identity and cooperation, we created three indices based on data from a random sample of 52 Chaldeans. The first index is a measure of the strength of a Chaldean's ethnic identity, (EI), the second is a measure of the extent to which a person reports cooperation with co-ethnics, and the third measures the extent to which the person reports cooperation with non-Chaldeans. The ethnic identity index includes the following variables: (1) ethnicity of people involved in the person's last three social activities; (2) ethnicity of the person's friends; (3) ethnicity of the person's spouse; (4) attendance at Chaldean churches, (5) whether the person considers Chaldeans to be Arabs, and (6) how much Chaldean is spoken in the home. Each of these variables, and all the variables in the indices used below, were normalized to have a mean of 0 and a variance of 1. Scores for each variable were averaged for each individual to create an over all measure for that person. The higher the value of EI, the greater the person's apparent ethnic identification.

The coethnic cooperation index (CC) consists of ten variables: five variables related to getting help in various situations, four variables related to getting professional services, and one variable measuring involvement in Chaldean organizations. Itemizing, these are: (1) where the person goes for help when he needs money for (a) business, (b) wedding, (c) funeral, and (d) medical bills; (2) where the person goes for help when he is sick; (3) the ethnicity of the person's (a) doctor, (b) lawyer, (c) dentist, and (e) accountant; and (4) whether the person is involved in Chaldean organizations. The greater a person's CC values, the more they tend to (or expect to) give and receive cooperation from other Chaldeans.

The index of cooperation with people outside the group (CNC) is made up of seven variables dealing with "public" cooperation (it's kind of a measure of civic-mindedness). These variables include: (1) voting, (2) recycling, (3) littering, (4) obeying stop signs, (5) working on a political campaign, (6) protesting a political issue, and (7) involvement in a non-Chaldean organization or group. The higher a person's score, the more he reports cooperating with non-Chaldeans. For details on the construction of these indices, please see appendix C.

Results from the Analysis

Our first step is to examine the relationship between our ethnic index (EI) and prosociality toward both Chaldeans (CC) and non-Chaldeans or civic coopera-tion (CNC). Figure 9.5 plots EI against CC and shows the kind of relationship anticipated by our coevolutionary theory of ethnicity. A simple regression shows that the standardized slope of the line in fig. 9.5 is 0.33 ($p = 0.018$). This implies that a standard deviation increase in our ethnic index predicts a one-third standard deviation increase in our Chaldean cooperation index.

Because the correlation between the indices could be confounded with other economic and demographic variables, which may be correlated with both ethnic identity and coethnic cooperation, we ran a series of multiple regressions to examine possible confounds. The variables examined included age, sex, house-hold income, and individual income. We did not find any of these variables to be significantly correlated with the indices, and even with all of these variables in the regression as predictors, the regression coefficient for EI was 0.35 ($p = 0.09$), which was three times greater than the next biggest coefficient.

Does being a high cooperator with coethnics correlate with generalized cooperation or civic-mindedness? Figure 9.6 uses our indices to show the

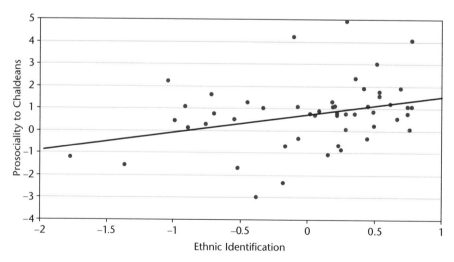

Figure 9.5 Ethnic identity and prosociality toward Chaldeans ($n = 51$).

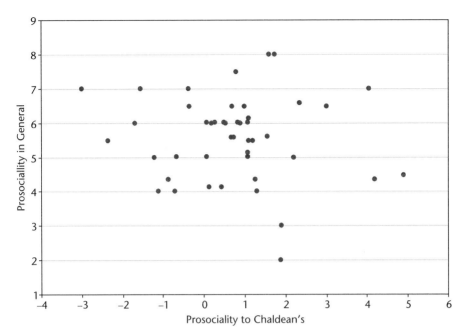

Figure 9.6 Relationship between prosociality toward Chaldeans and non-Chaldeans.

relationship between cooperation with coethnics and general prosociality. Although ethnicity and cooperation with coethnics are related, ethnic identity does not predict prosociality toward non-Chaldeans or civic-mindedness. There was no significant relationship between ethnic identity and cooperation with people outside the group. Furthermore, using the same demographic and economic variables as above to predict CNC, we find that none of them explain any significant portion of the variance in civic-mindedness, including EI.

Finally, there remains the question of how we may account for the tendency of some members of the Chaldean community to develop a greater degree of ethnic identification than other members. Our cultural-learning approach suggests that the community in which one grows up should have important effects on determining who one identifies with and how much one identifies. First, a simple regression analysis shows that an individual's age at arrival in the United States positively predicts her EI values ($\beta_{std} = 0.29, p = 0.05$). However, the effect of ontogeny should be highly nonlinear—that is, what should matter most is the first 20 or so years of life (Henrich forthcoming). To examine this, let's first lay out a simple table. Table 9.1 shows the mean EI values for those who arrive before age 20 and those who arrive after age 20. An analysis of variance shows a substantial effect ($p = 0.024$): Those who arrive before age 20 have much lower ethnic identification but a large variance (standard deviation). The relationship between age of arrival and EI can be seen in figure 9.7. These data suggest that some of those who arrived before adulthood may have

Table 9.1 Strength of Ethnic Identity by Age

Age	Mean EI value	Standard deviation	N
Before 20	−0.12	0.65	18
After 20	0.27	0.42	33

identified as Americans, or at least not as Chaldeans. If we run two simple regressions, one with the continuous variable "age of arrival" and a second with our "under 20 versus over 20" variable, we find that the simpler binary variable explains as much of the variance as the continuous one. Overall, this suggests that growing up longer in Detroit both decreases (on average) ethnic identification for "being Chaldean" and increases the variance in ethnic identification.

Summary

The measure that is best correlated with cooperation within the ethnic group is the strength of a person's ethnic identity. Though age, sex, number of years in the United States, income, number of people in household, and employment with relatives do not have a significant relationship with a person's likelihood of cooperating with coethnics, knowing how strongly a person identifies as a Chaldean is a good predictor of whether or not the person cooperates with,

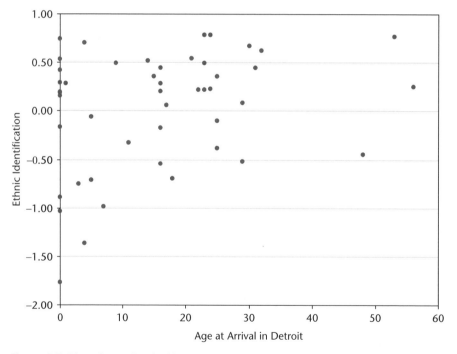

Figure 9.7 Plot of age of arrival in Detroit against Index of Ethnic Identification (EI).

and expects cooperation from, other Chaldeans. Although a person may be a good cooperator with other Chaldeans, this cooperative tendency does not reveal a generalized pattern of cooperation that extends beyond the ethnic group.

Conclusion

In this chapter, we have sketched a coevolutionary theory of ethnicity and examined Chaldean life through the lens of this theory. The theory is rooted in the simple idea that an individual can improve her payoffs if she can better coordinate her behavior with those with whom she is most likely to interact. In the deep background, genetic evolution created cultural learning as a means of rapid adaptation in information-poor environments (that is, environments in which it's difficult to figure out the adaptive thing to do). These cultural learning mechanisms tend to lead social groups to share a range of potential marker traits (e.g., dialect and dress) and underlying social norms because of how the mechanism happens to operate (e.g., conformist transmission) and because of the effects of social interaction on cultural learning. Natural selection, acting on genes, faces a social environment configured such that members of groups tend to have similar norms (e.g., rules for interacting in social exchange, such as a marriage or cooperative house building) and possess similar marker traits. Taking advantage of such regularities to more effectively coordinate exchange and avoid punishment, selection favored a psychology that preferentially (1) pays attention to and learns from individuals who display the same marker as oneself, (2) preferentially interacts with those showing the same markers, and (3) deploys essentialist reasoning to enable inferences about the underlying practices, beliefs, behaviors, and values of others when more specific information is lacking (i.e., people will tend to believe that people of the same ethnicity will all "naturally" have similar behaviors, beliefs, and values).

In addition to their role in the coevolutionary process, we also described how culture and cultural learning are important for understanding and explaining human behavior by, among many other things, defining the rules of ethnic inheritance and cues of group membership. Too, cultural practices can strengthen (e.g., ethnic endogamy) or mitigate (e.g., cross-ethnic adoption) the tendency to essentialize across certain ethnic lines. Thus, approaching ethnicity without an understanding of both genetic evolution and cultural evolution would leave one at a severe explanatory disadvantage.

Here we emphasize that ethnic groups, particularly successful minorities, will tend to construct cultural practices and institutions that maintain sharp boundaries, prescribe in-group endogamy, delineate religious or moral boundaries that are coterminous with the ethnic boundary, demarcate themselves linguistically, and foster competitive cooperation among coethnics. Groups that fail to do this tend to vanish via assimilation into larger groups. It's not that we don't think there are ethnic groups with traits inconsistent with these patterns; it's that such groups won't likely endure for centuries and millennia. This is a kind of cultural group selection, which we discussed in chapter 3. Though cultural group selection may favor these ethnicity-maintaining cultural traits, individuals' genetic self-interest may sometimes run in quite the opposite direction, particularly in the contemporary world.

Viewing the Chaldeans through this theoretical apparatus, we discussed how Chaldeans understand ethnic membership, their tendencies for preferential interaction and exchange with coethnics, and the relationship between ethnic affiliation and prosociality toward other Chaldeans. Consistent with our approach, our findings reveal that costly, hard-to-fake signals of group membership, specifically language and religion, are essential for full membership in the Chaldean community. These are the marker traits just mentioned that allow learners and potential members to bias their behavior. Here, costliness is important because if the markers are easy to acquire and fake, free-riders may enter the Chaldean system (lacking the required, though hidden, underlying values) and exploit the opportunities for cooperation. Along with these markers came a concern about "bloodlines" and/or phenotype, with many Chaldeans both refusing to bestow full membership to those without proper parentage and refusing to fully eject those of known Chaldean parentage who have rejected their church or other values. Indisputable Chaldean membership seemed to require all the costly signals and the proper parentage.

There is little doubt that Chaldeans preferentially interact, exchange, and cooperate with other Chaldeans. Despite being a respected minority scattered throughout a large metropolitan area, Chaldeans tend to seek out other Chaldeans for their professional needs, personal relationships, and commercial services. Moreover, they prefer to join organizations that help coethnics and strongly favor politicians principally on the basis of ethnicity. We sharpened this by showing that an individuals' degree of ethnic identification is positively associated with helping, and expecting help from, coethnics, although not with cooperation in general, or what might be called civic-mindedness.

Because much of what we have shown here should sound familiar to those who know the research on other minority ethnic groups, particularly in urban areas, we will briefly emphasize the nature of our contribution. Sociologists, anthropologists, and economists will no doubt note that there are alternative explanations for many of the phenomena we have observed among the Chaldeans. The strength of our approach here is not that it has revealed previously unobserved patterns, but rather that it begins to integrate sociological and economic approaches with psychological and evolutionary explanations. Our work derives all of this from a single theoretical framework, which we've already applied to other areas of human sociality and cooperation, and which has elsewhere been applied to an even broader range of questions. No other single theory we know of can explain (1) the essentialist tendencies in our psychology, (2) the consistent primacy of cues such as language and religion for ethnicity, (3) the consistent and sometimes individual costly biases in social interaction favoring coethnics, and (4) the relationship between ethnic identification and in-group prosociality. Even if there were such a theory, we would consider it as incomplete unless it also provided an account of ultimate origins. Only evolutionary theory allows us to ask the "why" questions.

10

Cooperative Dilemmas in the World Today

The philosophers have merely interpreted the world in various ways;
the point however is to change it.

—Karl Marx

Example is not the main thing in influencing others. It is the only thing.

—Albert Schweitzer

Studying cooperation and public goods can become highly theoretical, myopic, and abstract. It is easy to get bogged down in debating the assumptions of the mathematical models, the artificiality of the behavioral experiments, or the complexities of ethnographic research. In the midst of all the complexity needed to understand why and how humans have become a cooperative species, it is critical that we not lose sight of the urgent need to recognize and address the multiplicity of cooperation dilemmas embedded in many of the issues facing the world today, both locally and globally. Some of the most pressing health, economic, and environmental problems that we are currently confronting are, at their core, cooperation problems. Understanding the psychology of cooperation at the individual level and the repercussions of a person's, country's, or organization's failure to cooperate is necessary if we hope to improve our ability to tackle these issues. In this final chapter, we look briefly at a few of the most urgent cooperation problems that we (as a society, a nation, and a world) are grappling with, as well as a couple of examples of cooperation dilemmas that we face in our personal lives. These examples will illustrate how cooperative dilemmas permeate modern life.

Cooperation Is at the Core of Many Public Health Problems

The best weapon against disease is prevention, and one of the best ways to prevent disease is with vaccinations. Vaccination campaigns have been tremendously successful and can be credited for the control of many fatal and crippling illnesses. Perhaps the greatest success story of all is the complete eradication of smallpox in the 1970s via an aggressive vaccination campaign. The challenge of controlling infectious disease with vaccinations is that a critical mass must be vaccinated in order to keep the disease from spreading. The concept underlying vaccination is *herd immunity*. In order for a disease to spread, an infectious person has to come in contact with a person who is susceptible to catching and spreading the infection. If an infectious person comes in contact only with vaccinated people, then the disease can't spread. The sick person will either get better or die, and the contagion will cease to threaten others. If many, but not all, members of a population are vaccinated, then the chances that an infectious person will come in contact with a susceptible individual are very low, and thus the likelihood of the disease continuing to spread remains very low. The vaccinated people in the population act as a buffer for the unvaccinated people. But if too many members of a group decide to decline vaccinations, then herd immunity breaks down.[1]

Vaccination is a cooperation problem because there is some risk associated with vaccinations. In a small number of cases, a person may become ill or die from a vaccination. Using the paradigm laid out in chapters 2 and 3, we'd expect that once the percentage of vaccinated individuals gets sufficiently high, prestige-biased transmission and other forms of cultural learning should permit the spread of beliefs and practices that prompt some people to begin rejecting vaccinations (see example of polio vaccinations in Nigeria, below). This emerges in different ways in different places, but the core cooperative dilemma associated with herd immunity is the same. In North America, an increasing number of people are rejecting vaccinations for both religious reasons and because of medical myths (Public Health News Center, 2005). For example, there is a popular myth that thimerosal, a mercury-based preservative used in some vaccines, causes autism, and as a result, many people refuse to have their children vaccinated. The best available evidence does not support a causal relationship between thimerosal and autism, and thimerosal is no longer used in most vaccines[2] (Public Health Agency of Canada, 2004).

Even though getting vaccinated is the self-interested choice in many situations, social norms can stabilize any costly behavior, including rejection of vaccinations in a highly susceptible population (see chaps. 3 and 7). The effect of social norms on cooperating with vaccinations was evident in the 2005 outbreak of rubella in a series of religious communities in Ontario, Canada. At the center of the outbreak were members of the Dutch Reform Church, who shun vaccinations for religious reasons (Branswell 2005) and whose social networks consist largely of other church members. By the time the outbreak came under control, more than 120 people had been diagnosed with rubella. Although there were fears that the disease would spread, it remained relatively isolated. The containment of the outbreak can be attributed largely to high vaccination rates for rubella in the rest of Ontario, where it is estimated that 90 percent of the population is vaccinated, and to control measures put in place by the province. In this case, the religious beliefs of a population led many of its members to defect from vaccinating their children. The social networks of the group's members also played a critical role in the decision to not vaccinate, and on the impact of the non-cooperative behavior. The affected group lived adjacent to people who did not share their religion or their anti-vaccination beliefs, and this led to two distinct populations—one that vaccinated and one that did not. Consequently, rubella spread within the community that didn't cooperate with vaccination campaigns but did not spread in the community that cooperated.

A similar issue has emerged in Nigeria and neighboring countries, where polio had been nearly eradicated (except in one of the 36 Nigerian states, Kano). In 2003, some prestigious Islamic preachers in northern Nigeria (which includes the state of Kano) spread false rumors that polio vaccines were contaminated with antifertility agents and HIV. Many Nigerians subsequently refused to have their children vaccinated, and northern Nigerian politicians and health ministers temporarily banned the immunization campaigns in Kano and two other states (Cohen 2005). The rate of vaccination compliance in Kano dropped from a high of 80–90 percent to 15 percent in 2003. People who believed the rumors saw vaccination as having a very high personal cost that outweighed the risk of contracting polio. Once a group of individuals had refused vaccinations for their children, the number of susceptible people in Nigeria increased, and the few cases of existing polio were able to find new hosts. Polio made a comeback. It spread throughout Nigeria and into Ghana, Togo, Benin, Burkina Faso, Central African Republic, Chad, Cameroon, and Sudan (Altman 2004). At present, a strong campaign promoting polio vaccinations and publicizing the increasing risk of infection seems to be driving up vaccination rates (World Health Organization 2005). In Kano, the heart of the recent outbreak of polio, 92 percent of the children were vaccinated by late 2004, and the religious leaders who originally opposed the vaccinations have permitted their own people to be trained by health workers to give the vaccine (Linn 2005). Similar outbreaks of polio have occurred recently among unvaccinated Amish communities in Minnesota.

Eradication, or even just control, of many infectious diseases requires widespread vaccination of populations. Since diseases abide by no borders, the failure of one country to fully vaccinate its population leads to a potential snowball effect. If a country could achieve 100 percent vaccination of its population, then the disease could not reenter its population. But it is nearly impossible to vaccinate every individual as soon as he becomes susceptible[3] so populations rely on herd immunity. However, when an adjacent population has many susceptible people, then the disease can become sufficiently common to penetrate the herd immunity. To have global eradication, as is the goal for polio, all countries must cooperate with the vaccination campaign and agree to incur the minimal individual risks in order to achieve the greater good of eliminating a crippling disease from the world.

The general success of vaccination campaigns in the United States also exemplifies how punishment can be used to encourage cooperative behavior. This is the strategy most commonly used to enforce vaccination in the United States, where children cannot attend public schools unless they provide proof that they have received all of their childhood vaccinations. This has been a successful method of motivating parents to vaccinate their children, although some parents who oppose vaccinations are immune to this threatened punishment because they homeschool their children or obtain immunization exemptions for religious or medical reasons. Because vaccination (or lack of it) presents a cooperative dilemma in a large population, theory shows that the only way to sustain cooperation and prevent both the invasion of defecting strategies (non-immunizers) and infectious diseases is through formal or informal sanctions. Perversely, giving the public more (true) information could result in less compliance, since this would reduce the power of conformist transmission (see chap. 2) to stabilize cooperation and the punishment of noncompliance, and it would clarify the self-interested choice when herd immunity is high. Free-riding, even in light of knowledge about herd immunity, could be avoided if people have adopted a strong value or preferences to actions that benefit the group over the individual and his kin.

Cooperation as a Political Problem

The Global Fund is an international health funding mechanism that raises, manages, and distributes money to fight HIV/AIDS, malaria, and tuberculosis. The funding comes from governments and private sectors from around the world, and the funds are distributed to experts and local communities for prevention or treatment. As of 2005, more than 45 countries and other donors had pledged $6.1 billion to be paid to the fund through 2008 (Global Fund 2005). However, collecting the full amount pledged has become a cooperation problem.

In 2004, the United States set aside $547 million to donate to the fund that year. But contributing the full amount was contingent on donations being made by other countries, with the United States stipulating that it would not provide more than 33 percent of total donations. It would donate all $547 million only if other donors contributed a total of $1.1 billion. Actual donations were $243 million short, so the United States planned to withhold $120 million (U.S. Mission to the United Nations in Geneva 2004; CBS News 2004). In the end, the United States contributed nearly $459 million in 2004, approximately $88 million less than the amount set aside for its annual donation (Global Fund 2005). With so many countries pledging money to the fund, it is tempting for donors to withhold some of their pledged money. But if one country holds back, then other countries feel that they are shouldering an unfair portion of the financial burden, and they may respond by decreasing their donations. As the number of countries and the amount withheld increase, a snowball effect is set in motion, and the gap between pledges and donations grows.

Though it could have been the case that each donor country would see its contribution as independent, in reality many donors viewed the pool of money in the global fund as a public good, and donation decisions were linked to the contributions of the other donors; this was most explicit for the United States, which officially caps it contributions based on total contributions. Consequently, pledge fulfillment is at risk of defections, which can cause a cascade effect and leave the fund well below its target. The position of the United States could have negative repercussions that extend beyond the money that it doesn't contribute. Because the United States is the only superpower in the world and has great prestige, other countries may model their donor behavior on that of the United States, as we've described in the process of prestige-biased transmission (chap. 2). If everyone copies the U.S. strategy, there would be a downward spiral in donations until no one was paying anything.

The U.S. policy is remarkably similar to some of the strategies associated with the social norms psychology we discussed in chapter 7, or what has been called *strong reciprocity* by Fehr, Bowles, Gintis, and colleagues, and clearly documented in Public Goods Games. The United States is invoking a rule of contingent cooperation.[4] This, however, is not a smart move for a prestigious leader. As discussed in chapter 2, prestigious individuals are preferentially imitated. Thus, a prestigious leader, whose goal is to maximize cooperation, should adopt a strategy of altruism and punishment. This means the United States should lead by example, giving generously and punishing those who don't give by damaging their reputations. This kind of leadership will cause some to imitate our generosity and some to imitate generosity and punishment (which will induce more cooperation from those who would otherwise defect). This takes us back to chapter 3, where we showed that the core dilemma of cooperation involves creating a correlation between being a cooperator and interacting with other cooperators.

Cooperation as an Environmental Problem

The most obvious cooperation problems are those involving the use and protection of the environment and natural resources. Everyone uses and is affected by the environment. The effects of the actions of any one individual are insignificant, and the cumulative effects of the actions of many individuals are enormous. All people use environmental resources, and some environmental resources, such as air, are used by every person on earth. When billions of people are all using the same common resource, a cooperative dilemma is created that can only be solved if leaders step forward and organize, regulate, or inspire people to cooperate. Because so many people use the same resources, many individuals believe that they should not control their own use of the resource because other people aren't controlling their use. If one person limits his use, but others do not, then he will be incurring an individual cost even though the benefit of his regulated behavior will have no effect on the overall sustainability or health of the resource so long as everyone else in the world continues to use the resource unabated.

One way in which the global community has come together to try to address the problem of global warming is with the Kyoto Protocol. This international agreement includes a legally binding framework for the reduction of greenhouse gas emissions by industrialized nations by 2012. The agreement went into effect on February 16, 2005, shortly after it was ratified by Russia. The agreement could go into effect only when it was ratified by 55 nations, including those industrialized nations that account for 55 percent of all industrialized nations' 1990 carbon dioxide emissions (Environmental Literacy Council 2005). Because of these conditions, ratification by Russia was critical; although 140 countries had committed, industrialized nations were underrepresented, and ratification by a major industrial polluter was needed. The United States had already made very clear that under the Bush administration, the United States would not ratify. The American refusal to participate jeopardized the entire treaty because the United States is the world's largest emitter of carbon dioxide from the burning of fossil fuels (CIA 2005; see fig. 10.1 for total carbon dioxide emissions in select industrialized nations); without its signature, it would be difficult to satisfy the conditions for the agreement to go into effect. Furthermore, there was fear that the American refusal would result in a domino effect, prompting other countries to back out. Other nations could adopt the attitude common to cooperation problems: "If the United States isn't going to reduce emissions, why should we? Besides, if the United States doesn't cut back emissions, nothing we do will really make a difference. It has to be all or none to make this work!" Fortunately, the science and facts of global warming were persuasive enough that other countries refused to let the U.S. defection jeopardize a program that could play a significant role in repairing and protecting the environment.

Without some kind of norm (in this case, an international contract), reducing emissions is a classic public goods problem. The costs of reducing emissions are borne by the individual country, but any benefits generated are distributed

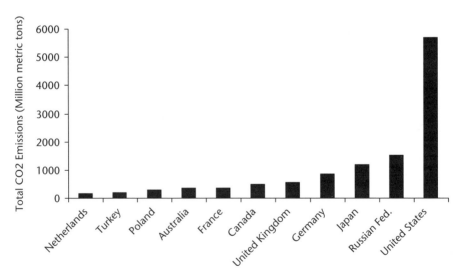

Figure 10.1 Total carbon-dioxide emissions (in millions of metric tons) in a sample of industrialized nations for 2001.

Source: World Resources Institute 2005.

worldwide. Given this, what led to the successful ratification of the Kyoto Protocol? Three of the factors discussed throughout this book contributed to the initial success of this cooperative dilemma.

First, the agreement itself restructured the cooperative dilemma into a step-level public goods problem in a manner seemingly designed for conditional cooperators of the kind discussed in chapter 7. Nations that agreed to cooperate did not have to pay the costs of cooperation unless at least 55 other nations also agreed to pay the costs. At the margin, this structure had a crucial psychological effect on Russia. In holding back, Russia became the linchpin. If it signed, the treaty would go into effect; if it did not, the agreement would fail, at least for the immediate future. As the last signatory, Russia was empowered to deliver a potentially big benefit.

Second, the Kyoto Protocol recognized the importance of punishment for cooperation. Unlike the 1992 United Nations Framework Convention on Climate Change, which was the predecessor to the Kyoto Protocol, the Kyoto Protocol included legally binding commitments rather than voluntary commitments. Without the threat of punishment, a few defectors can drive conditional cooperators to full defection—a fact repeatedly demonstrated in Public Good Games. Thus, the test will come for the treaty if and when certain countries do not live up to their obligation under it. If punishment is not administered to defectors, and the punishment clauses in the treaty are revealed to be toothless, conditional cooperators will begin to waiver (perhaps just defecting a little), and the Kyoto Protocol may unravel.

Third, within some countries, substantial portions of the population have adopted culturally transmitted beliefs and values related to the environment, making conservation and population regulation a valued goal in and of itself. In democratic countries, the values of these constituencies helped guarantee early signatories to the agreement (which are important, based on our studies of conformity) and ensured that at least some nations will stick to the agreement and sanction those who do not (even if the official sanctions turn out to be somewhat toothless). As we showed in chapters 7 and 8, some people stick to a norm, including a cooperative norm, regardless of the threat of punishment.

The interplay of these three factors helped turn the Kyoto Protocol into a success story (as to ratification) in which 141 countries came together and agreed to incur the necessary costs to work collectively to protect the global environment, despite the defection of the world's largest source of fossil-fuel-related carbon dioxide emissions (Note: U.S. fossil-fuel emissions are 60 percent higher than those of the world's second-largest emitter, the People's Republic of China, and represent 24 percent of global emissions; see Marland, Boden, and Andres 2005).

Cooperation as a Workplace Problem

Cooperation problems can occur wherever more than one person is interacting, such as in the workplace. Anyone who works outside the home is familiar with the problem of employees' coming to work when they are sick. There are many reasons why you may go to work sick, even though you risk infecting others. Among the reasons: (1) using up sick days that could be saved for times when you're even sicker or that could be used to take a day off (why waste a sick day on staying home sick when you can use it to take a long weekend?); (2) losing income, in the form of wages, lost sales, or business; or (3) falling behind in work that still needs to get done. When you stay home because of sickness, you may incur costs, but your coworkers benefit by continuing to enjoy a healthy workplace. However, the workplace retains a healthy environment only if people stay home when they're sick. If one person defects and comes to work ill, then the healthy status of the workplace declines. Other people may decide that they too will come to work sick because the workplace is already unhealthy, so there's no reason to stay home or because they don't want to incur the costs of staying home when other people aren't willing to do the same. Workplaces develop a culture or set of norms about coming to work when sick. Though some offices are able to maintain their healthy status, other workplaces accept that people may come to work sick.

Reducing illness in the workplace can be achieved by restructuring some of the costs associated with staying home, by changing employee and management attitudes about working when sick, and by creating sick-day policies that encourage sick workers to stay home. A complication to solving the workplace problem is that people can be infectious before they have symptoms and thus may inadvertently harm the health of the workplace. This would be an unintentional defection. To avoid this accidental defection, coworkers would have to be willing to accept that accidents happen and to give the offender another chance by seeing if he stays home the next time he's sick. This is analogous to applying some of the logic from the generous and contrite TFT strategies we discussed in chapter 3.

The workplace situation is similar to one we find in some pediatricians' offices. Some pediatricians have two waiting rooms—one for healthy children who are there for such things as annual checkups, and one for sick children. If everyone takes his children to the appropriate waiting room, then healthy children won't get sick from going to the doctor's office. But there is an incentive for parents bringing a sick child to the doctor to defect and go to the healthy waiting room. Even though my child may be sick, I still don't want her to catch the illnesses of the other children in the sick waiting room. If I take my sick baby to the healthy waiting room, I protect my child from getting any additional illnesses, even though I am exposing healthy children to my child's illness. As a protective parent who is primarily concerned about the health of my own child, I have a strong incentive to defect and abuse the healthy waiting room. But if everyone did this, then the healthy waiting room would become as unhealthy as the sick waiting room, and the public good would be destroyed for everyone, including my own child. The next time I'd bring her for a well-child checkup, I would no longer have a healthy room in which to wait.

Proper use of the waiting rooms is done on the honor system. However, other factors help to maintain the status of the healthy waiting room. First, many parents seem to have strongly internalized the norm that it is wrong to intentionally expose other children to your child's illness (chaps. 7 and 8). Second, other parents will be angry if they realize that you have brought a sick child into the healthy waiting room—and it will be pretty obvious if your child coughs or sneezes, or has a runny nose. The other parents may deter the offender with dirty looks or by reporting the offender to the receptionist. Either of these would be embarrassing, and the risk of being caught could deter some potential defectors (chap. 7). So people are most likely to cheat if they have a sick child who appears healthy. Third, pediatricians' offices are often used by parents from the surrounding community, and they may know each other. A parent's reputation can be hurt if someone she knows sees her in the healthy waiting room with a sick child. Risk of getting a reputation as a person who puts other children's health in jeopardy may deter defections (chap. 6). Though all of these factors may collectively contribute to the maintenance of the healthy waiting room, we suspect that the first probably plays the largest role.

Cooperation Problems Are All around Us

As social beings who live and work together and who share and compete for limited resources, we constantly encounter, and sometimes create, cooperation problems. At times we cooperate out of habit, without even realizing it. Other times, lacking a social norm, we miss opportunities for cooperation. The scale of the problem can range from one involving two people that can easily be solved with reciprocity, to a global problem involving every person on the planet that can be solved only with regulations, carefully constructed incentive structures, monitoring and enforcement, and the evolution of appropriate social norms. The challenge of solving these problems is twofold. First, the situation has to be recognized as a public goods problem. Second, the complexity of solving a cooperation problem has to be appreciated. There are few cases that can be solved with only one strategy. In most cases, different people will be motivated by different strategies, and the most effective way to achieve cooperation is by employing multiple approaches, such as education about the rightness of behaving a certain way coupled with punishment for not acting that way. The Kyoto Protocol presents a good example of how a complicated cooperative dilemma was addressed by restructuring the problem, building in punishment and reputational consequences, and relying on countries where many people have internalized appropriate norms. Cooperation problems are complex, and often their solutions demand an integrated understanding of our evolved social psychology, the importance of existing cultural values and history, and the dynamics of cultural transmission and social norm formation. The complexity of the solutions is not a reason to give up on the problem. Rather, it gives hope that they can be solved and cooperation expanded and sustained.

Appendix A: The Underlying Structure of Cooperation

Joseph Henrich

All genetic evolutionary explanations of cooperation are successful to the degree that they allow natural selection to operate on statistically reliable patterns or regularities in the environment. All too often, the assumptions or constraints that maintain this regularity are submerged in the basic structure or setup of the model and are not explicitly analyzed. To expose the required regularity, I develop a very simple model of cooperation. We begin with a population of N individuals indexed by i and subdivided into a number of smaller groups labeled with j. Each individual possesses either an altruistic gene (altruistic = cooperative, for purposes of this exposition) or an egoistic gene tracked by x_i. When $x_i = 1$, individual i possesses the altruistic allele (version of the gene), and $x_i = 0$ marks i as possessing the egoistic allele. To derive the conditions for the spread of an altruistic gene, we will use the famous Price Equation (G. Price 1970, 1972):

$$\overline{w}\Delta\overline{x} = Cov(w_i,x_i) + E(w_i\Delta x_i) \tag{1}$$

The Price Equation is a simple statistical statement that relates the expected change in the frequency of a gene ($\Delta\overline{x}$) per generation, the absolute fitness ($w_i = $ *the* number of offspring of i), the average fitness of the population ($\overline{w}$), and the current frequency of the gene in i (x_i). The $Cov(w_i,x_i)$ term gives the covariance between absolute fitness and gene frequency, and $E(w_i\Delta x_i)$ represents the expected value of the within-individual change in x_i (e.g., mutation) weighted by the individual's absolute fitness. Following Hamilton (1975), I am using the nonstandard notational convention of leaving the subscripts on the variables inside of the covariance and expectation operators.[1] In general, equation (1) provides a very general statement about any evolutionary system.

In this model, altruists will bestow benefits on other members of their local group j at a cost to themselves. Egoists will not bestow benefits on others, but both egoistic and altruistic group members will benefit from being in groups with many altruists. To investigate this, I specify w_i as follows:

$$w_i = a + \beta_{w_i x_i \bullet x_j} x_i + \beta_{w_i x_j \bullet x_i} x_j + \varepsilon_i \tag{2}$$

Here i's fitness is jointly determined by the effect of her genotype on her fitness, holding her local group composition constant, $\beta_{w_i x_i \bullet x_j} x_i$ (x_j is the average value of x_i in group j), and the effect of i's local group on her fitness holding the individual's genotype constant $\beta_{w_ix_j \bullet x_i}x_j$. The constant baseline fitness is α and ε_i is the uncorrelated error term.

With equations (1) and (2), we can derive a general statement about the conditions for the genetic evolution of altruism (a propensity to cooperate). Because we are interested only in the effects of natural selection, I will ignore nonselective forces such as recombination, drift, meiotic drive, and mutation. This means $\Delta x_i = 0$, so the expectation term drops out of (1). Substituting (2) into (1) yields

$$\overline{w}\Delta\overline{x} = Cov(\alpha,x_i) + \beta_{w_ix_i} Var(x) + \beta_{w_ix_j}\beta_{x_jx_i} Var(x) + Cov(\varepsilon_i,x_i) \qquad (3)$$

By definition, $Cov(\alpha,x_i)$ and $Cov(\varepsilon_i,x_i)$ are zero. Setting $\Delta\overline{x} > 0$ gives the conditions for an increase in the frequency of the altruistic gene in the population (assuming both alleles exist in the population):

$$\beta_{w_ix_i} + \beta_{w_ix_j}\beta_{x_jx_i} > 0 \qquad (4)$$

Equation (4) shows that all solutions to the evolution of altruism—whether they are based on kinship, reciprocity or group selection—are successful according to the degree to which "being an cooperator" predicts that one's partners or group members are also cooperative (Frank 1998). By definition, $\beta_{w_ix_i}$ is negative because, *ceteris paribus*, altruists have lower fitness than egoists; $\beta_{w_ix_j}$ is always positive because, independent of one's own genotype, it is always better to be in a group with more cooperators. This leaves $\beta_{x_jx_i}$, which captures the degree to which "being an altruist" predicts "being in a cooperative group." If groups are randomly remixed every generation, $\beta_{x_jx_i} = 0$, and altruism (even kin-based cooperation) cannot evolve. However, if (for whatever reason) cooperators can preferentially group with other altruists, $\beta_{x_jx_i}$ will be a positive number between 0 and 1, and cooperation at least has a chance. If egoists can preferentially enter groups with altruists, $\beta_{x_jx_i}$ will be negative. Thus, altruistic genes can spread only if $\beta_{x_jx_i}$ is sufficiently close to one such that the benefits to being in an altruistic group outweigh the costs of bestowing benefits on others. Those with any familiarity with evolutionary theory may recognize (4) as a generalized version of Hamilton's Rule (Queller 1992). Expressing this in the standard notation, $\beta_{w_ix_j}$ is b, the fitness benefits provided by altruists; $\beta_{w_ix_i}$ is $-c$, the cost to the altruist; and $\beta_{x_jx_i}$ is r, which in the specific case of kin selection is interpreted as the coefficient of relatedness by descent from a recent ancestor. In general, however, $\beta_{x_jx_i}$ is *not* a measure of relatedness by descent from a recent common ancestor—although that is *one way* to get a positive value of $\beta_{x_jx_i}$ Hereafter, for simplicity, I will refer to $\beta_{x_jx_i}$ as β.

Most important, equation (4) shows that understanding particular solutions to the evolution of altruism requires analyzing how the model maintains a sufficiently high value of β. Thus, the acceptability or legitimacy of a theoretical solution depends on an evaluation of the constraints (or "special assumptions") that give rise to, and maintain, the statistical association (β). Remember, however, that the greater the value of β, the greater the amount of altruism that can evolve *and* the greater the selective pressures for mutant genes that can beat the system by exploiting β. Under different circumstances, in different species, different kinds of constraints or special assumptions may be warranted. The details of an organism's social structure, physiology, genome, cognitive abilities, migration patterns or imitative abilities may support some hypotheses and undermine others. Of course, by emphasizing the species specificity of certain constraints, I do not mean to suggest that each constraint is equally likely to be observed in a randomly selected species. Surely, some constraints are more frequently satisfied in nature than others; nevertheless, the rarely satisfied constraints may provide the most interesting forms of altruism. Unfortunately, despite the need to defend a particular constraint vis-à-vis the details of that species that might justify focusing on that constraint over others, many students of evolution and human behavior regard only kinship- and reciprocity-based models of altruism as legitimate solutions. In contrast, approaches based on quite plausible constraints that may be satisfied only in the human case, such as those based on cultural group selection and culture-gene interaction, are neither widely considered nor understood by nonspecialists. For a more extensive discussion along these lines, see Henrich 2004a.

Appendix B: Ethnographic Research Methods and Challenges

Natalie Henrich

This appendix describes the methodology for the ethnographic research with the Chaldeans of metropolitan Detroit and unforeseen difficulties that arose during the study, which I conducted from September 1999 through December 2000. Fortunately, the complications, though at times frustrating, led to additional insight into the culture.

Research Methods with the Chaldeans

I began my research by conducting interviews with randomly selected members of the community. My goal was to acquire information, gain familiarity with the culture, and explore avenues for more informal interaction with Chaldeans. My first challenge was recruiting interview subjects. Unlike conducting research in remote, homogeneous villages where the presence of a researcher is an interesting and novel event, locating Chaldeans in a large urban center is quite a feat. Luckily, the Chaldean church publishes a Chaldean directory every few years, and I used the 1998 edition of this directory as my primary source for interview subjects. Because the Chaldean community in metropolitan Detroit is very large (approximately 100,000 people) and divided economically (the new immigrants and the poor live in Detroit, whereas the more affluent Chaldeans live in the suburbs surrounding Detroit), I decided to narrow the research population to Chaldeans who live in Southfield, a middle- to upper-class suburb on the northwest side of Detroit. Many Chaldeans live there, and it has, among other things, the largest Chaldean church (in terms of number of congregants), a Chaldean social club (the Southfield Manor), the only Chaldean retirement home, the office of an umbrella organization for most Chaldean groups (the Chaldean Federation of America), and a popular Chaldean-owned restaurant (La Fendi). Although the Southfield Chaldeans were the focus of my study, I ended up expanding the research population to include a small number of Chaldeans from other Detroit suburbs of comparable socioeconomic circumstances. The reason for this is discussed below.

After deciding to concentrate on Chaldeans from Southfield, I set out to make a list of random individuals who could be interviewed. I did this by going through every page of the Chaldean Directory[1] and selecting the first person on each page who lived in Southfield; there may have been individuals from Southfield who were not selected because their home address was not included in the directory. If there were no Southfield residents on a page, I moved on to the next page. For each entry, the directory includes all or some of the following information: the person's name, home phone number, home address, place of employment, work phone number, work address. Once I compiled my random list of Chaldeans I did two things. First, if the person's occupation was party store owner,[2] I would go to his store, ask for the person from my list, introduce myself and ask if I could conduct an interview. In almost every instance, the person agreed to be interviewed. Depending on how busy the store was, and how much time the person had, I either conducted a full interview or asked a subset of questions. Most of the stores have a bulletproof barrier between the employees and the customers. In all but two stores, I was invited behind the barrier, and we had the interview off to the side, in an office, or by the cash register, with the grocer answering questions between helping customers. There were several grocers whom I continued to visit periodically. We developed a friendship and engaged in a "gift-giving" relationship. I would bring cookies and cakes, and they would give me Middle Eastern desserts, stews, meat pies, and so on. One man in particular took me under his wing and introduced me to Chaldean foods, taught me some of the Chaldean language, and always welcomed my presence warmly.

Second, for people who were not store owners, I would call up the individual, introduce myself, and ask if I could meet with the person for an interview. As with the grocers, almost everyone with whom I spoke agreed to be interviewed. We arranged to meet in their homes, at coffee shops and restaurants, at their offices, at public libraries, or in the Mother of God church (the Chaldean church in Southfield). With very few exceptions, all the people were welcoming and helpful beyond the call of duty, offering to speak with me again and to help in any way possible. When we met at coffee shops or restaurants, the person I'd called often treated me to a coffee or a meal, despite my attempts to pay. And when I met with people in their homes, I was usually served tea and cookies. The Chaldeans I studied are generally affluent and proud and would not accept money for their participation, so I would follow up with flowers or pastries as a way of thanking them for their help. The interview session was often very sociable, with chatting both before and after the interview.

In addition to contacting people through the Chaldean Directory, I also asked the people I interviewed if they knew anyone else who might agree to be part of the study. Most people said no, but a few did give me names and numbers of friends and relatives. This was one of the ways that I ended up including non-Southfield Chaldeans in my sample population.

Another way that I made contacts in the community was through Mother of God church. For the first couple of months of my research I attended English mass each Sunday at the church.[3] Although I did not meet anyone at mass, it did provide me the opportunity to observe people as they interacted before and after the service, to see how they dress, and to learn more about their religion (Catholic, Chaldean rite) and traditions. At the end of mass, people made announcements, and in my first week there an announcement was made about a group of young adults that meets at the church. Later that week, when I was speaking with one of the priests regarding hiring a Chaldean translator and research assistant, I asked him about this group. He invited me to attend its weekly meeting, and thus I was introduced to the CARE group (Chaldean Americans Reaching and Encouraging). The group consists of Chaldeans, primarily in their twenties, who work to improve the Chaldean community through undertakings such as conducting food drives for the Chaldeans in Detroit, visiting the elderly in nursing homes, mentoring Chaldeans in high school, raising money for scholarships, and so on. Through my involvement with CARE, I developed friendships, learned about the goings-on of the community and became more integrated into the community. I also interviewed several CARE members and some of their parents—this was another source of non-Southfield interview subjects.

An additional way that I got to know the community and the culture, and to give something back to the community, was by volunteering at the Community Education Center. This is a Chaldean-owned, government-funded, free education center. Although it is open to anyone, it is located on the edge of Chaldean Town in Detroit, and all of the teachers and most of the students are Chaldean. I taught an introductory computer class, which gave me regular interaction with the Chaldean staff and students. Throughout the year, I also attended various community meetings and social events, in addition to weekly CARE meetings.

My primary data comes from my work with the interview subjects. Most subjects were interviewed twice, between three and ten months apart. Interviews generally lasted between 45 minutes and an hour, although a few lasted between two and three hours. In some of these cases, the extended duration of the interview was a result of Chaldeans' giving very long answers and using the interview as an opportunity for introspection and reflection on their community. In the other cases, the interview portion of the encounter was the typical length but then the interview became a social visit with snacks, home videos, chatting, and, in one case, singing.

After the interview, each subject was solicited to participate in two laboratory-style economic experiments, one during each of the interviews. These experiments are described in chapter 8.

Interview-Related Research Challenges

The first difficulty encountered was in finding a research assistant/translator. With the strong economy and tight labor market, anyone who spoke English and wanted a job already had employment. Consequently, I was able to conduct only two interviews with non-English-speaking Chaldeans. In one of these interviews, I was assisted by a member of CARE who volunteered to attend the interview with me and serve as a translator. In the other case, the daughter of the person I was interviewing translated during the interview. Fortunately, most non-English speakers live in Detroit (as opposed to the suburbs) and thus were not part of my research population. Within my population, the non-English speakers tended to be elderly, but even among the elderly there were many people who spoke English. Consequently, I was able to conduct interviews with this segment of the population although there may be some bias in the data because non-English-speaking elderly people are underrepresented.

The next challenge is one common to field researchers. Often, people would schedule an interview with me but fail to show up. Although this was frustrating and added to the time it took me to conduct the interview portion of my research, it did not bias my sample because I rescheduled almost all of the missed interviews and did eventually interview all of these people. This means that the data are not biased toward people who kept their initial appointments.

Unfortunately, not everyone who did the first interview was available for a second interview. Approximately half of the people who participated in a first interview did not do a follow-up interview or the second experiment because of changes in phone numbers and e-mail addresses that prevented me from contacting them, or because the person was not interested in participating in a second interview. Several of the grocers whom I recruited as subjects by showing up at their stores were not asked for a second interview because I felt as if I had imposed on them the first time and was not comfortable asking for more of their time. There were also two grocers with whom I became friends, and I did not ask them for a second interview because the nature of our relationship had changed. My impression was that they felt that now that we were friends, I shouldn't think of them as informants, so my visits were strictly social, with the occasional question tossed in casually. This last reason for not being able to conduct some of the second interviews highlights the positive experience I had of working with the Chaldean community. The people were warm, friendly, and forthcoming, and I formed relationships that were both highly informative and personally valuable.

Experiment-Related Research Challenges

Conducting experiments with subjects who were randomly selected from a nonstudent population created an unexpected difficulty: fifteen percent of my initial informant pool did not want to participate in the games. Most Chaldeans didn't understand how the experiments could be contributing to an understanding of their culture, and this led some people to decline to participate. I couldn't provide them with an explanation of how the experiments would help me understand their culture prior to playing the games because it could bias their behavior in the games. Some people wouldn't play without this information, so I explained everything to them, but then didn't let them do the experiments.[4]

Several people didn't want to do the experiments because they were uncomfortable with using money as the medium for payoffs (other people expressed similar concerns but still agreed to participate). Three people said that the games sound like gambling and that they don't gamble, and one man seemed to think that I was somehow profiting from the experiments. Others said that they only accept money that they have worked for and that they couldn't accept money for playing the game because they hadn't earned it. These people were men and women over the age of 40. These informants said that they didn't need the money, and I think that they perceived the payoffs from the game as charity. To accept a handout would be an affront to their work ethic and would imply that I thought they needed money. A 70-year-old man played the UG but when I came back 9 months later for his second interview and told him about the DG, he got visibly upset and said that he regretted that he had played the UG because he should never have accepted money he hadn't worked for and that he didn't want to participate in the DG. Other people made statements such as "I don't want to take any money, I'm not interested in being wealthy. I have money to meet my needs and don't need more," "I have enough money," "I won't take money unless I earn it. I won't gamble, and I need to deserve and sweat for what I get," and "I won't accept anything I don't work for. Money that comes easily, goes easily." One man said that he would play if we could use his $20 rather than my research money and suggested that I play the game only with poor people who need the money. People who did play also expressed this sentiment; they thought that the experiment was a good way to get money to people in need. A few people said that they would play if they didn't have to accept any money (play the games hypothetically) or if we played for something else, like fruit or cookies. Money turned out to be a culturally loaded medium.

The man's offer to use his money rather than my research funds also reflected another point of confusion in the experiments. I had to explain several times to most players that the MacArthur Foundation and NSF provided the money for the experiment, and that it was not my personal money on the line. This was important to clarify because some people at first didn't want to participate because they didn't want to take my money away from me, and others thought that if they rejected an offer, then they were helping me because I would get to keep the money. During the explanation of the rules, I had to make this point very clear and to tell them that if they reject the offer, I take the money and play with other people until all the money is used up.

Some of these challenges may arise when working with other cultural groups, since other researchers have reported similar difficulties. Gil-White (2004) found that Mongols often didn't understand that the money came from a funding agency rather than Gil-White's personal funds. Consequently, his players rejected offers so that Gil-White could keep his money. Gowdy, Iorgulescu, and Onyeiwu (2003) found that subjects in a Nigerian village were suspicious of the source of the money, suspecting that the money was obtained from magic related to human sacrifice. Fortunately, a member of their research team was a former resident of the village, and he convincingly explained that the money was not tainted.

When working with nonstudent populations, experimenters should be prepared for confusion about the purpose of the experiments because the games appear very strange and abstract to people who are unfamiliar with academic research. Careful attention should also be paid to cultural interpretations of terms, the medium of the stakes, and how the context of the game is perceived.

Appendix C: Constructing the Ethnicity and Cooperation Indices

To investigate the relationship between ethnic identity and cooperation, we created three indices, which are based on data from 52 individuals. The first index is a measure of the strength of a Chaldean's ethnic identity, the second is a measure of the extent to which a person cooperates with coethnics, and the third measures the extent to which the person cooperates with people outside the group. A difficulty arose in creating the ethnic identity and the coethnic cooperation indices because some of the variables could be classified as indicative of either identity or cooperation. For example, if a Chaldean goes to a coethnic lawyer, doctor, accountant, and dentist, does this indicate a strong ethnic identity or a tendency to cooperate with coethnics? In the end, we decided that anything that could be cooperation would go into the cooperation index, leaving only purely identity variables in the ethnic identity index. As a result, the ethnic identity index includes the following variables: (1) ethnicity of people involved in the person's last three social activities; (2) ethnicity of the person's friends; (3) ethnicity of the person's spouse; (4) attendance at Chaldean church, (5) whether the person considers Chaldean to be Arabs, and (6) how much Chaldean is spoken in the home. Each variable was assigned a value, which was then normalized to have a mean of 0 and a variance of 1. For each individual, we took an average of her scores across each variable, and this became her score for the index. The higher the person's score, the stronger is her ethnic identity. Looking at each variable, the breakdown is as follows:

1. *Ethnicity of people from last three social events*: For the ethnicity of people with whom the person socialized during his last three visits, a person could receive a value between 0 and 3. If none of the events was with Chaldeans, then the person received 0; if one event was with Chaldeans, he received 1, and so on.
2. *Ethnicity of friends*: The ethnicities of a person's friends were assigned scores between −2 and 2. If a person has no Chaldean friends, he received −2; if he has a few Chaldean friends but mostly non-Chaldean friends, he got −1; if he has equal number of Chaldean and non-Chaldean friends, he got 0; if he has mostly Chaldean friends, he got 1; and if he had all Chaldean friends, he received 2.
3. *Ethnicity of spouse*: If the person is married, or engaged, to a Chaldean, she got 1. If the person is married, or engaged, to a non-Chaldean, she got −1. If the person was unmarried, she got 0.

4. *Attendance at Chaldean church*: If a person reports attending Chaldean church regularly, he got 1; if he attends on occasion, he got 0; and if he never attends, he got −1.
5. *Chaldeans as Arabs*: If a person considers Chaldeans to be Arabs, she got 0; and if she does not consider Chaldeans to be Arabs, she got 1.
6. *How much Chaldean is spoken in the home*: Individuals were assigned a 0 for no Chaldean spoken in the home, a 1 for some, and a 2 if people reported speaking mostly or entirely Chaldean in the home.

The coethnic cooperation index includes ten variables, consisting of five variables related to getting help in various situations, four variables related to getting professional service, and one variable measuring involvement in Chaldean organizations. The specific variables are: (1) where people go for help when they need money for business, wedding, funeral, or medical bills; (2) where people go for help when they are sick; (3) the ethnicity of their doctor, lawyer, dentist, and accountant; and (4) whether they are involved in Chaldean organizations. As in the ethnicity index, each variable received a value that was normalized, and the higher a person's score, the more he cooperates within the Chaldean community. Looking at each variable, the breakdown is the following:

1. *Getting financial help for business*: If a person would never go to another Chaldean for help, he got −2. If he would go first to a non-Chaldean source, such as a bank, but would ask a Chaldean friend for help as a last resort, he got −1. If he was equally likely to go to a non-Chaldean or Chaldean source, he got 0; and so on. Note that we did not include going to a relative for help as going to a Chaldean because this could be considered kin selection rather than a consequence of ethnic identity. Only institutions and unrelated individuals (Chaldean or non-Chaldean) were considered in the assignment of the variable's value. This criterion applies to all variables involving getting help.
2. *Getting financial help for weddings, funerals, and medical bills*: If the person would go to a non-Chaldean source for help, she got −1. If she would go to either a Chaldean or non-Chaldean source, she got 0; and if she went exclusively to Chaldeans for help then she got a 1. Going to a relative for help did not count as going to a Chaldean.
3. *Ethnicity of one's lawyer, doctor, dentist, and accountant*: For each of these variables, if the person goes only to non-Chaldeans then he received −1; if he goes to both Chaldeans and non-Chaldeans he got 0; and if he goes only to Chaldeans, he got 1.
4. *Involvement in Chaldean organization*: If the person was involved in at least one Chaldean group or organization in the last year, she received a 1. If she was not involved in any Chaldean groups or organizations in the last year, she received a 0.

The index of cooperation with people outside the group is made up of seven variables dealing with "public" cooperation. These variables include: (1) voting,

(2) recycling, (3) littering, (4) obeying stop signs, (5) working on a non-Chaldean political campaign, (6) protesting a non-Chaldean political issue, and (7) involvement in a non-Chaldean organization or group. For each variable, a score between 0 and 1 was assigned. The higher a person's score, the more she cooperates with non-Chaldeans. Looking at each variable, the breakdown is as follows:

1. *Voting*: If the person voted in the last local election, she scored 1. If she did not vote, she scored 0. The same applies for participation in the last state and national elections.
2. *Recycling*: If the person recycles, he got 1. If he doesn't recycle, he got 0.
3. *Littering*: If the person never or rarely litters, she got 1. If she litters, she got 0.
4. *Obeying stop signs*: If the person always stops at stop signs, he got 1. If he ever runs stop signs, he received 0.
5. *Political campaigning*: If the person has ever worked on a non-Chaldean political campaign, she got 1. If she has never worked on a campaign, she got 0.
6. *Protesting*: If the person has ever protested a non-Chaldean political issue, he got 1. If he has never protested an issue, he got 0.
7. *Involvement in a non-Chaldean organization/group*: If the person was involved in at least one non-Chaldean organization or group in the previous year, she got 1. If she has not been involved in a non-Chaldean organization or group in the previous year, she got 0.

Reliability of Indices

The data used in the indices come solely from self-reports. Consequently, this raises the issue of whether people with a strong ethnic identity are more likely to overreport cooperation with coethnics, thus creating a false correlation between ethnic identity and coethnic cooperation. Ideally, we would test reported prosociality with an independent measure. Lacking such a measure, we structured the interview questions so as to minimize the opportunity for this type of bias. For example, rather than asking, "Do you use any professional services from Chaldeans?" respondents were asked, "Is your dentist Chaldean or non-Chaldean?" "Is your lawyer Chaldean or non-Chaldean?" and so on. Similarly, we did not ask people if they have worked on a campaign for a Chaldean candidate. Instead, respondents were asked if they have ever worked on any political campaign, and if they did, they were asked to name all the candidates they have helped. Also, many questions had follow-ups that increased the accuracy and robustness of the answers. For instance, Natalie would ask if the person works with any relatives, and if he said yes, Natalie asked him to list all of the people in terms of their relationship to him (i.e., son, brother, wife). The follow-up questions allowed us to confirm that the person was currently working with relatives and to determine whether the coworker was true or fictive kin. In gathering data for the ethnic identity index, we also tried to frame questions so as to minimize subjectivity. We asked specific questions, such as "Is your spouse/fiancé Chaldean?" and "How many of the last three visits were with Chaldeans?" (rather than, say, "Do you usually visit with Chaldeans?").

Notes

Chapter 1

1 Both Darwin and Baldwin suspected that culture might influence genetic evolution. Baldwin called it "social heredity" (Baldwin 1896), and Darwin referred to inherited "habits" (Darwin 1981: chap. 5).

2 To protect our Chaldean informants, we have altered any identifying references (names, obvious occupations, etc.) to individuals in a manner that preserves the ethnographic depth without jeopardizing the promise of confidentially made at the commencement of interviews.

Chapter 2

1 In the early 1930s, Wintrop and Luella Kellogg (1933) reared their 10.5-month-old son, Donald, with a female chimpanzee named Gua, whom they adopted at age 7.5 months. Donald immediately began imitating Gua's behaviors—see the main text. Gua, however, seemed to acquire little (via imitation) from Donald or his parents.

2 In thinking about content biases, it is important to keep in mind two things. First, evolutionary products, like human minds, are likely to contain accidental by-products and latent structures that create biases for fitness-neutral behaviors, ideas, beliefs, and values. Boyer (2001) details one kind of by-product content bias in his explanation for the universality of some religious concepts (e.g., ghosts). Second, even content biases that arose because they led to the adoption of fitness-enhancing behavior in ancient environments may now promote the adoption of quite maladaptive practices.

3 In social situations, a person's payoffs and those of others are jointly influenced by the choices of all involved. In nonsocial situations, only the decision maker's payoffs are influenced by his own decisions.

4 Regarding the incentives for performance, readers from business schools should not be concerned that the doubled-digit dollar amounts typically used in experimental games might not motivate MBA students. Knowing their students, the experimenters put something that MBAs really care about on the line: their grades. Class credit was assigned according to a participant's position relative to the median performance in the experiment, with the top performer getting an additional 20 percent added to her final grade, and the bottom performer losing 15 percent.

5 Payment was determined by randomly picking among these subround decisions such that the total number of paid rounds was equal between the control and the two treatments.

6 Methodological problems of earlier research were avoided here (Asch 1948).

7 The attractiveness construct shows very little consistency across researchers, and it doesn't always involve varying *physical* attractiveness while controlling for other sources of likeability (Petty and Wegener 1998:345).

8 Kameda and Nakanishi (2002) include a nice experiment that confirms both the existence of conformist transmission and the predicted effects of making individual learning more costly.

9 Presence of the model during the donation phase of the experiment had little effect (Rosenhan and White 1967), although the experiment described below had the model leave the room (while the experimenter observed the subject, alone in the room, through a one-way mirror).

10 When children observe a model that says one thing (e.g., "Give a lot") and does another (gives nothing), the children seem to mostly ignore the model's words and attend primarily to his actions in deciding how much to give. As a result, children who observed a model give nothing donated, on average, less than the baseline amount given by children in the control group (Bryan, Redfield, and Mader 1971; Midlarsky, Bryan, and Brickman 1973).

11 Children will even imitate some specific self-critical comments made by the model (Herbert, Gelfand, and Hartmann 1969).

12 Natural selection seems to have acted on different genes to produce the same effect (extended lactase production) among the African versus European populations (Mulcare et al. 2004; Scrimshaw and Murray 1988).

Chapter 3

1 Analyses of the probability of being caught cheating on income taxes (e.g., for false deductions or unreported income) and the sizes of the penalties show that most people pay too much (Skinner and Slemrod, 1985), if all they care about is money. Since punishment is insufficient, "not cheating" becomes a contribution to the public good.

2 Note that votes for Al Gore exceeded those for G. W. Bush by more than a half million nationwide.

3 We realize that there is some debate about whether recycling, when all the costs and effects are taken into account, will actually achieve these beneficial ends. However, as will be come clear below, and in chapter 5, social norms can potentially maintain any behavior, even ones that don't yield overall benefits.

4 Those who doubt the group benefit or cooperative nature of orderly lines have not spent much time in countries where it is not the custom to make orderly lines for scarce resources.

5 This assumes that these "costs" and "benefits" translate in some manner (however weakly) into differences in survival and reproduction.

6 Economics has also sought to explain the puzzle of cooperation. Interestingly, however, the reasons why the puzzle has arisen in economics has more to do with the disciplinary tradition of assuming that individuals are self-interested rather than with any prima facie deductive logic. Other than its heritage in Enlightenment philosophy, there is no reason (that we know of) why economics has typically assumed pure self-interest. Adam Smith (2000), for example, wrote eloquently about the importance of moral sentiments—see the quotation opening this chapter.

7 We can rule out genetic evolution for these changes in social behavior because the time periods of these historical changes are too short.

8 Dawkins (1976) first presented this colorful "green-beard" example.

9 "Greenbeard solutions" continually reemerge in the literature on the evolution of cooperation. Unknowingly, researchers seem to produce "solutions" to the evolution of cooperation that are actually green-beards in disguise. Having set up green-beards as a nonviable solution to cooperation, it must be noted that there is at least some evidence for green-beards in nature (Keller and Ross 1998).

10 Key theoretical work on kin selection comes from Hamilton (1964, 1972), Grafen (1985), Frank (1997), and Queller (1992).

11 It is a common misconception that kinship depends on sharing a certain *percent* of the same genes, and that r gives the fraction of genes shared by two relatives. This is false for two reasons. First, natural selection will favor genes that direct benefits at identical copies of themselves, not other genes, so percent of shared genes is theoretically irrelevant. Thus, r should be thought of as the probability that another individual has a copy of the "helping gene" given that the first individual has it. Second, humans already share most of the same genes because we are the same species. Sharing genes for building fingers and blood vessels is not important for understanding human cooperation. Thus, r does not give the percent of shared genes; it does happen to correspond to the percent of genes that are identical by descent from a recent common ancestor. But, as we said, this is only relevant in that it may create reliable patterns that natural selection can exploit.

12 The power of human imitation causes people who live in close proximity to look similar. Because people are continually (and unconsciously) mimicking each other's facial expressions and body positions, facial musculature and ethological patterns cause those living in proximity to resemble each other (Zajonc et al. 1987). Similarly, the more time people spend together, the more likely they are to acquire patterns of dress, body decoration (makeup), and hairstyle, all of which contribute to phenotypic similarity. Eating from the same hearth (similar diet and spices) can create similar body odors.

13 Kin selection actually offers some explanatory power here regarding the role of the mother's brother, although it falls far short of eliminating the need for a complementary cultural explanation. The thing about fathers is that they can rarely be completely positive that the offspring of their wife (or any woman they've had sex with) is definitely their progeny. In contrast, a mother and all of her relatives can be quite sure of the relatedness of her child (putting aside artificial insemination). From the perspective of a father, if the probability that a child is, in fact, his progeny drops below 50 percent, then the mother's brother is, on average, a closer male relative. Although keenly interesting, this argument does not deal with why some societies have low paternity certainty, or why some societies have norms regarding the role of the mother's brother, which are enforced regardless of the paternity certainty in any *particular* case. That is, even when the paternity of a particular child is known, the mother's brother's still plays the fatherly role. Our coevolutionary explanations of social norms can provide the missing pieces to this argument.

14 Readers interested in exploring all the ways in which TFT fails should begin with Bendor, Kramer, and Stout 1991; Bendor, Kramer, and Swistak 1996; Boyd 1989, 1992; Boyd and Lorderbaum 1987; and Hirschleifer and Martinez 1988.

15 The relevant theoretical findings were independently arrived at by different researchers around the same time, using somewhat different models (Bendor and Mookherjee 1987; Boyd and Richerson 1988; Joshi 1987).

16 All of Axelrod's early work was in a "noiseless environment" in which strategies always had perfect information about their partners' past behavior.

17 These insights were gleaned from a variety of sources (Bendor 1993; Bendor, Kramer, and Stout 1991; Bendor and Swistak 1997; Boyd 1989; Hirshleifer and Martinez 1988; and Wu and Axelrod 1995). Another interesting aspect of generosity in noisy environments is that the most successful strategies overall may lose to *every* other strategy in pairwise competition. In contrast, a strategy that defeats every other strategy in pairwise competition may place dead last when all the strategies are let lose together. To understand this, suppose a strategy LOSER was paired with nine other strategies for 10 rounds each, and LOSER lost by a score of 9 to 10 to eight of these other strategies, and by a score of 1 to 2 to the ninth strategy called VIG (vigilant). This would give LOSER a total of 73 points out of 90 possible points. Further, suppose that VIG defeated all the other strategies by a score of 2 to 1 in dyadic competition, giving VIG a total of 18 points. Thus, if the world contains a mix of the 10 different strategies, LOSER will often be the overall winner because LOSER can induce most other strategies into cooperation. NICE (cooperates initially) and GENEROUS guys don't always finish last.

18 Those familiar with evolutionary psychology might note that this is the same "combinatorial explosion" logic used to defend the theoretical claim that our cognition is composed of many special-purpose modules. Thus, the same logic that leads one to believe the mind is modular also leads one to be rather suspicious of arguments that humans have "reciprocity psychology" that operates *without* adaptive cultural input.

19 We think the intuitive appeal of TFT clouded what should have been an evident puzzle: if TFT is so simple and robust, why don't we see more cooperation of this type in nature? Answer: TFT isn't that robust.

20 Most of human adaptations to our diversity of physical environments are cultural. Blowguns, kayaks, boomerangs, bone tools, and poison arrows are all adaptations in the classic sense, and they are almost entirely culturally learned (Henrich, forthcoming).

21 The TFT-like strategies are easily derived from imitation. It is simply "cooperate initially" and imitate the other guy. Young children readily engage in reciprocal imitative exchanges of all kinds (Meltzoff and Prinz 2002). Thus, imitation provides a solid foundation for the emergence of reciprocity.

22 Any reading of these papers by Nowak and Sigmund should be accompanied by a reading of Leimar and Hammerstein (2001), who provide an important critique.

23 With the exception of Boyd and Richerson 1989, serious empirical and experimental work did not begin until 1998, perhaps with the publication of Nowak and Sigmund's (1998b) paper in the journal *Nature*. Alexander (1987) coined the term *indirect reciprocity*, but did not formally model it.

24 When a cooperative unit has more than two people (i.e., more than two people are contributing to the provision of a shared good, such as taxes or a restaurant bill), it is impossible to withhold cooperation from a defector without also withholding it from cooperators, and this causes all the cooperation in the group to unravel. Here's why: When a cooperator decides not to cooperate in response to a defection, he withholds cooperation not only from the person who previously defected but also from all the people who cooperated. The remaining cooperators then perceive the punisher as a free-rider, so they respond by withholding cooperation from him. This process repeats itself until all the reciprocators stop cooperating.

25 For anyone familiar with ethnography or history, this would be an obvious modeling project. Mathematical modeling should be motivated by such empirical cases.

26 Biases such as self-similarity, sex, age, and ethnicity can further hone the accuracy of one's reputational information. For detailed models showing how conformist transmission can improve noisy information, see Henrich and Boyd 2002.

27 We believe that this form of indirect reciprocity is equivalent to forms of costly signaling in which individuals are signaling their value as future cooperative players. There are other forms of costly signaling that could lead to cooperative acts that are not part of indirect reciprocity (Bliege Bird, Smith, and Bird 2001; E. Smith and Bliege Bird 2000).

28 Some students of evolution and human behavior have a visceral reaction to anything that uses the term "group selection." Fear not. Cultural group selection models do *not* carry the problems typically associated with models of genetic group selection (Boyd et al. 2003; Henrich 2004a; Henrich and Boyd, 2001).

29 In analyzing the evolution of cooperation, the real challenge is to explain the maintenance or stability of cooperative and prosocial (punishment of noncooperators) behavior or tendencies, not their spread from zero. All kinds of random effects, population shocks, stochastic migration patterns, and so on to the evolving system can make cooperators common at a particular place and time. Thus, the key is to explain why these cooperators won't simply vanish back into evolutionary history as the system returns to equilibrium. The same situation holds true for reciprocity solutions; TFT, for example, is not favored until it is sufficiently common.

30 Boyd et al. (2003) take advantage of this logic in a computer simulation to show that, even without conformist transmission, the introduction of punishing strategies substantially enhances the effect of competition between social groups, and thereby favors the evolution of substantial amounts of pro-group cooperation.

31 It is important to realize that indirect reciprocity alone does not predict this kind of punishment. First, indirect reciprocity only works for cooperative behaviors, and as we'll show, people punish norm violations even when the norms are noncooperative (e.g., violating a food taboo or greeting custom). Second, indirect reciprocity does not predict punishment; it only predicts disengagement or defection. Unlike defection, punishment reduces the payoffs of both parties.

32 In each period of a mutual aid game, one randomly selected player from a group of N is "needy," and the other N-1 players can "donate" (at a cost to themselves) or not. If the needy person has a good reputation, then those who donate get (or keep) a good reputation while those that do not get a bad reputation. If the needy person has a bad reputation, donating or not donating does not influence one's reputation.

33 It is possible that the mutual aid game can be used to maintain cooperation in multiple public goods games. By creating cooperation in the truth-telling game, the fidelity of reputation can be maintained sufficiently to allow the mutual aid game to also stabilize cooperation in another *n*-person game, such as warfare or conservation.

34 Note the parallel with our discussion of kin selection, where we explained how smell or hanging around mom can provide reliable predictors of relatedness, and thus reliable predictors on having the same gene.

35 We do not have the space here to deal with this at any length, but this theory of ethnicity explains the essentialist and primordial nature of human thinking vis-à-vis race as a by-product of the coevolutionary processes that produced it for ethnicity (Gil-White, 1999, 2001). In human ancestral history, individuals probably rarely, if ever, encountered individuals who looked very different from themselves (i.e., in terms of skin color, morphology, hair, etc.). When people did finally meet such individuals, these phenotypic differences, by chance, fit easily with the cues for different ethnic groups; racial difference may superstimulate this tendency to "species thinking." To be perfectly clear, we are *not* saying that there are essential or primordial differences between either racial or ethnic groups. We are addressing why, given that there are not such differences, humans so frequently think about such groups in these terms.

Chapter 4

1 Seventy-six percent of Chaldeans interviewed said that a Chaldean who leaves the Catholic Church is still a Chaldean. Catholicism is therefore not a requirement for being a Chaldean.

2 This historical information was compiled from conversations and interviews with typical members of the Chaldean community, Chaldean priests who are experts on Chaldean history, and informational packets and books written by members of the community. Although most members of the community have at least a basic knowledge of the history presented here, not everyone knows as much detail about particular historical figures and dates as the priests and the more historically educated Chaldeans.

3 Natalie was told by several Chaldeans that the Mesopotamians and Babylonians were all Chaldeans, and thus one should study the history of the Mesopotamians and Babylonians to learn the history of the Chaldeans. Consequently, many historical figures who may not be known as Chaldeans by non-Chaldeans are in fact considered such by Chaldeans—Hammurabi provides the most common example. However, other Chaldeans distinguish earlier occupants of this region to be ancestors rather than full-fledged Chaldeans.

4 Most Chaldeans lived in the northern Iraqi village of Telkaif.

5 Chaldeans interviewed were randomly selected from the Chaldean directory, but the sample from which individuals were drawn was restricted to people living in Southfield. Additional Chaldeans from comparable suburbs were introduced into the sample as Natalie met members of the community at church events and meetings of charitable organizations and was subsequently introduced to friends and colleagues (see Ethnographic Research Methods, Appendix B).

6 Neither of us knew the woman being interviewed because she had been selected as part of Natalie's random sampling of the community. For details of our ethnographic methods, see Ethnographic Research Methods in Appendix B.

7 Keep in mind that our sample includes only Chaldeans living in the suburbs of Detroit, with the majority living in Southfield—a middle- to upper-class area. The estimated median household income in Southfield in 2000 was $61,315, with 22.5 percent of households having an income of $100,000 or more. Estimates come from income data for Southfield in 1989, reported in the 1990 U.S. census, which we increased by 51.1 percent. The rate of increase was determined from the U.S. Department of Housing and Urban Development's (2000) median family income for the state of Michigan in 1989 and 2000.

8 At the time of the research, Detroit only had small grocery stores. None of the major chains had stores within the city, although they did have stores in the surrounding suburbs.

9 We doubt that grocers made these individual attributions to impress the ethnographer; in other contexts, Chaldeans readily made (what in our view were) blatantly racist statements without any hesitation or understanding of how such statements might be understood by the ethnographer. Examples are provided in later chapters.

10 This rate of 28 percent is based on data given in Putnam (2000: 59–60). He reports that in 1987, 50 percent of Americans reported that they were active in an organization. In the same year, 14 percent reported having a leadership role in an organization. In 1994, only 8 percent of people reported having a leadership role. Assuming that the rate at which active members hold leadership roles has remained constant over time, we extrapolated that the percentage of active members in 1994 had declined to 28 percent.

Chapter 5

1 This has not stopped people like the renowned cultural anthropologist Marshall Sahlins (1977) from attempting to argue that human kinship and kin-related behaviors have nothing to do with genealogical relations, as prescribed by Hamilton's Rule (see chap. 3). Sahlins's own examples were later shown to provide excellent evidence for the explanatory power of evolutionary theory (Silk 1980).

2 Laboratory studies with undergraduates suggest that matrilineal grandparents invest more in their grandchildren than do patrilineal grandparents (Gaulin, McBurney, and Brakeman-Wartell 1997).

3 We lack the space to explain why this first claim is dubious. Suffice it to say that the inferences which lead one to the conclusion that humans lived in small stable groups of close kin is based on a quite superficial analysis of the extant foraging and horticultural groups. There is good reason to believe that these groups are poor representatives of Paleolithic societies (Fehr and Henrich, 2003).

4 Although people do not generally mistake all (or most) nonkin in their groups as kin, there are situations in which people are consistently erroneous in identifying kin. Several well-known studies on incest aversion reveal that humans develop incest aversions to nonrelatives if they are raised together as children (Fessler and Navarrete 2004; Shepher, 1983; Wolf, 1995). This implies that humans may have a psychology that causes us to develop feelings of kinship toward people with whom we grow up even when they are not biologically related.

5 It should be pointed out that our kin psychology probably has not evolved to take into account interbreeding (cousin marriage among grandparents, for example), so from a psychological view it is probably irrelevant.

6 Lending money qualifies as cooperation because the lender incurs the costs of giving up access to his money while the borrower has it and forgoes interest on the money (the loans are usually interest free). There is also the risk that the loan will never be repaid, in which case the cost of the loan can be substantial. On the other hand, the borrower receives a great benefit from the loan, especially when he pays no interest.

7 Relying on kin as one's only employees is, eventually, impractical as a business grows and needs large numbers of employees.

8 The difficulties of hiring nonkin (especially non-Chaldeans) are recognized by many: several former store owners explained that they had to sell their stores when their siblings or children decided to work in other careers because they could not run the store without their help (family is the foundation for the stores, though some owners hire additional nonkin employees). The cost of hiring nonkin employees was likely one factor in selling off the stores, although issues of trust and a preference to work with family may also have contributed to the decision to sell.

Chapter 6

1 Engelmann and Fischbacher adapted their design from Seinen and Schram.

2 This situation could also be constructed as a mutual aid game, which can be solved by reputation-based strategies (Sugden 1986).

3 A skeptic might suggest that individuals are inclined to lie and say that they gave more, when they actually gave less. We have several reasons for doubting this interpretation. First, Natalie generally asked initially, "What do most people do?" and then eased into "What do you do?" Interviewing experience and supporting results from psychology suggest that people will assume that what they do or believe is common. So,

for example, cheaters tend to believe that cheating is more prevalent than it actually is. So although people may be disinclined to admit their own stinginess, they would likely (often with enthusiasm) project this onto others. Second, most people said they gave equally, and thus avoided an opportunity to lie and make themselves look good. Third, Chaldeans were generally amazingly forthright in their opinions, as people openly stated racist opinions about blacks and expressed admiration for (and racism against) for Jews.

4 It is also important that the bridesmaids include the cost of their dresses in their gifts because the bride buys the dresses.

Chapter 7

1 Much of the material in this chapter has been discussed in the literature under the label "strong reciprocity," although our precise interpretation is somewhat different from other authors. We have avoided using the term because, in our earlier drafts, novice readers were confused between our discussion of reciprocity (as in chap. 4) and strong reciprocity. In terms of both their empirical (and psychological implications) and evolutionary foundation, strong reciprocity differs substantially from (direct or indirect) reciprocity.

2 Prestige-biased cultural group selection: because humans have a tendency to learn from individuals who are more successful (higher payoff), ideas, beliefs, values, and practices will tend to spread from groups with greater economic production, health, and wealth, to groups with less, merely by differential cultural learning.

3 The pattern shown in figure 7.1 is not dependent on placing the punishment portion of the experiment after the nonpunishment portion. Fehr and Gächter (2000) show that putting the punishment round first does not substantially affect their findings.

4 Here, we've presented only the tip of the iceberg. There is a substantial literature on the effect of punishment on cooperation in laboratory experiments (Masclet et al. 2003; Ostrom, Gardner, and Walker 1994; Yamagishi, 1986, 1988).

5 One argument that Cosmides and colleagues might use to try to counter this theoretical observation is to claim that human psychology is not geared to understanding the anonymity of modern life because group sizes in ancestral environments were small and people did not encounter strangers sufficiently often in fitness-relevant ways. There are two problems with this. First, the effect of group size in killing reciprocity is dramatic (Boyd and Richerson 1988). The chances for reciprocity to produce much cooperation as group size increases above 10 get vanishingly small. Even the smallest forager groups have ten adults. Second, the available data on ancestral environments, which is rather thin, suggests that ephemeral contacts with strangers were probably common (or at least not rare exceptions), and likely were fitness-relevant (Fehr and Henrich 2003). Claims to the contrary rely principally on widely believed and frequently repeated anthropological myths rather than on the available data from ethnographically and historically known foragers, paleoarchaeological studies of ancestral human populations, and field research on nonhuman primates.

6 We are not suggesting that costs and benefits do not matter. Instead, we are predicting that cultural salience (having culturally learned something as a norm) will have a big impact on when and where people cooperate, punish, and stick to certain behaviors.

7 The Panchanathan and Boyd (2004) model, which expands indirect reciprocity by putting it into a cultural evolutionary framework, could also explain some of this. In their model, reputation can stabilize costly cooperative behavior in public goods

dilemmas, but there is no costly punishment (see chapter 2). Some have misunderstood this model as dealing with genetic evolution (Henrich 2006).

8 In cooperative situations, economists have also termed this *altruistic punishment* (Fehr and Fischbacher, 2003; Fehr and Gächter, 2002). It should be noted that under some circumstances this could be explained by the Boyd and Panchanathan model, in which the "punishment" is a defection against the person being punished, such that the defector ("punisher") yields net benefits rather than costs.

9 Recycling and recycling bins were common in metro Detroit at this time.

Chapter 8

1 A description of methodological challenges and difficulties that arose from conducting the experiments in a nonstudent population is presented in appendix B.

2 Approximately 85 percent of the people Natalie initially interviewed participated in the experiments. A few additional subjects were recruited outside the interview context.

3 This stake was chosen because, at most, only a very small effect from stake size had been found in previous experiments (Camerer 2003), and so there was no justification for using larger stakes.

4 Not all UG players participated in the DG, but all DG players had previously played the UG.

5 There were two exceptions to this. In both of these cases, the subjects played the UG and the DG on the same day. However, they played the DG first, and then they were responders in the UG and were given offers of 50 percent. At this stage of the research, we were very confident that a 50-50 UG offer would not be rejected, so we didn't worry that participating in the DG would contaminate a player's decision to accept or reject the UG offer.

6 With the outlier included, $p = 0.17$ (MW).

7 Most economists never do postgame inquiries. If they do, they use impersonal questionnaires, rather than semistructured interviews.

8 In our experimental work with the Machiguenga (Henrich and Smith 2004), we found that people also expressed the notion that being assigned the role of proposer or responder was just luck. However, instead of using that to suggest an equitable division, the Machiguenga felt that the responder should take whatever he could get—Machiguenga accept very low offers.

9 It is interesting to note that although almost all of the dictators gave 50 percent of the money to the recipient, no one gave more than half. As in real-life situations, people are generous and helpful, but not to the point that they are worse off than the other person. We suspect that if we had announced the DG offers at church, dictators would have mostly given it all.

10 For example, see Fiske 1992.

Chapter 9

1 Yamagishi and Yamagishi (1994, 1998) explain cooperative within-group behavior in experiments using these same ideas: expectations of shared ideas and strategies among group members reduce the risk involved in interacting and can lead to increased cooperation. They do not apply this to an evolutionary explanation for within-group cooperation but do use it to account for much of the observed behavior in experiments.

2 Gil-White does *not* argue that ethnic groups actually do have an "essence" (on the contrary). Rather, he argues that people tend to think about them that way.

3 At present, most people do still speak Chaldean, but the younger generations speak among themselves in English and older people say that younger Chaldeans do not speak the language as well as their parents and grandparents.

4 See Nettle and Dunbar (1997) for a discussion of language as a social marker and its ability to promote cooperation within a language group.

5 People did not talk about a "genetic inheritance" but rather talked about being Chaldean by "blood." It seems that some Chaldeans believe that there is some Chaldean essence that gets transmitted from parent to offspring, although the vehicle of transmission is ambiguous. We don't know if people would say that there is a Chaldean gene. However, a professional man in his forties said that the financial success of the store owners may be genetic. Trying to figure out the secret of the success of the Chaldeans, the man said, "There must be something genetic about that group coming from northern Iraq [Chaldeans from the Telkaif region]. If he buys a grocery store, he ends up buying a supermarket. They all do very well. It must be genetic. It's definitely hard work, and a desire not to have a boss. But it's something more than that."

6 These data contrast somewhat with Gil-White's (2001) adoption study among the Mongols.

7 Some Chaldeans consider Assyrians and Chaldeans to be the same ethnicity, but it was recorded as non-Chaldean when the respondent classified the friend in this way.

8 These individuals list themselves in the Chaldean Directory and/or participate in Chaldean organizations. Note, however, that economic motives exist for wanting to be in the Chaldean Directory, even if you aren't all that "Chaldean." Moreover, though using the Chaldean Directory does add some bias to our data, this method is substantially more random and systematic than are approaches typically used in urban ethnographies.

9 Based on a total population in Wayne County of approximately 2,061,162 (U.S. Census 2000) and an estimated 100,000 Chaldeans.

Chapter 10

1 When herd immunity breaks down, the risk of being exposed to a contagion is greater than the risks associated with getting vaccinated. With this change in the costs and benefits of vaccination, it becomes in an individual's best interest to get vaccinated and protect herself directly. At that time, there is no longer a cooperation problem because individual and group interests are aligned. But as more people become vaccinated, then the incentive to free-ride off of herd immunity kicks in. This creates a threshold for the percentage of the population that is vaccinated in which the situation shifts from one in which individual costs and benefits favor vaccination to one in which they don't.

2 Thimerosal is still used in the flu vaccine, although alternative stabilizers are being developed for use in flu vaccines for children.

3 Infants are born with immunity to some diseases and do not become susceptible and in need of immunization until later in infancy.

4 Anyone who thinks that this looks like reciprocal altruism needs to return to chapter 3. Reciprocal altruism sharply predicts defection (not contingent cooperation) as the size of the cooperative group grows to more than two individuals.

Appendix A

1 For example, using a more standard notational convention, one could write $Cov(w,x)$ in lieu of $Cov(w_i,x_i)$.

Appendix B

1 The directory includes 329 pages of Chaldeans living in and around Detroit. It also includes a subdirectory of Chaldeans from outside of Michigan.

2 The type of small grocery stores typically owned by the Chaldeans are locally referred to as "party stores."

3 Mass is held three times each Sunday in English, Arabic, and Chaldean (Aramaic).

4 It should be noted that typical laboratory economic experiments are not even a random sample of university students, let alone a random sample of "Americans." University subjects are often recruited from paid subject databases (which are signed up for voluntarily) or by using signs advertising the chance to make money.

References

Alexander, Richard. 1987. *The Biology of Moral Systems.* New York: Aldine De Gruyter.

Alpesteguia, Jose, Steffen Huck, and Jorg Oeschssler. 2005. Imitation: Theory and Experimental Evidence. *Governance and the Efficiency of Economic Systems*, Discussion Paper 54. Available at: http://www.gesy.uni-mannheim.de/dipa/54.pdf. Accessed October 24, 2006.

Altman, Lawrence. 2004. Spread of Polio in Africa Makes UN Fear a Major Epidemic. *Global Policy Forum.* Available at: http://www.globalpolicy.org/socecon/develop/africa/2004/0623polio.htm. Accessed October 24, 2006.

Asch, S. E. 1948. The Doctrine of Suggestion, Prestige, and Imitation in Social Psychology. *Psychological Review* 41:258–90.

Atran, Scott, Douglas Medin, Norbert Ross, Elizabeth Lynch, Valentina Vapnarsky, Edilberto Ucan Ek', John Coley, Christopher Timura, Michael Baran, Fracisco Gil-White, Nora Haenn, Giyoo Hatano, Keiko Takahashi, Ueli Hostettler, and Nora Haen. 2002. Folkecology, Cultural Epidemiology, and the Spirit of the Commons: A Garden Experiment in the Maya Lowlands, 1991–2001. *Current Anthropology* 43(3):421–450.

Atran, Scott, Douglas Medin, Norbert Ross, Elizabeth Lynch, John Coley, Edilberto Ucan Ek', and Valentina Vapnarsky. 1999. Folkecology and Commons Management in the Maya Lowlands. *Proceedings of the National Academy of Sciences.* 96:7598–7603.

Axelrod, Robert. 1984. *The Evolution of Cooperation.* New York: Basic Books.

Axelrod, Robert, and Lisa D'Ambrosio. 1994. *Annotated Bibliography on The Evolution of Cooperation.* Ann Arbor. Available at: http://www-personal.umich.edu/~axe/research/Evol_of_Coop_Bibliography.htm.

Axelrod, Robert, and William Hamilton. 1981. The Evolution of Cooperation. *Science* 211:1390–1396.

Baksh, Michael George. 1984. *Cultural Ecology and Change of the Machiguenga Indians of the Peruvian Amazon.* Unpublished dissertation. University of California at Los Angeles.

Baldwin, James Mark. 1896a. A New Factor in Evolution. *American Naturalist* 30:441–457, 536–554.

———1896b. Physical and Social Heredity. *American Naturalist* 30:422–428.

———1968. *Mental Development in the Child and the Race: Methods and Processes.* New York: Augustus M. Kelley.

Bandura, Albert, and Carol J. Kupers. 1964. Transmission of Patterns of Self-Reinforcement through Modeling. *Journal of Abnormal & Social Psychology* 69(1):1–9.

Baron, Robert. 1970. Attraction Toward the Model and Model's Competence as Determinants of Adult Imitative Behavior. *Journal of Personality and Social Psychology* 14:345–351.

Baron, Robert, Joseph Vandello, and Bethany Brunsman. 1996. The Forgotten Variable in Conformity Research: Impact of Task Importance on Social Influence. *Journal of Personality & Social Psychology* 71(5):915–927.

Bayless, T. M., and N. Rosenwei. 1966. A Racial Difference in Incidence of Lactase Deficiency. *Journal of the American Medical Association* 197(12):968–972.

Bazzi, Michael. 1998. Who are the Chaldeans? In *1998 Chaldean Directory*, pp. 24–34. Detroit, MI: Cal Press.

Bearman, Peter. 2004. Suicide and Friendship among American Adolescents. *American Journal of Public Health* 94(1):89–95.

Beckerman, Stephen, and Paul Valentine, eds. 2002. *Cultures of Multiple Fathers: The Theory and Practice of Partible Paternity in Lowland South America*. Gainesville: University of Florida Press.

Beja-Pereira, A., et al. 2003. Gene-Culture Coevolution between Cattle Milk Protein Genes and Human Lactase Genes. *Nature Genetics* 35(4):311–313.

Bendor, Jonathan. 1987. In Good Times and Bad: Reciprocity in an Uncertain World. *American Journal of Political Science* 31:531–558.

——— 1993. Uncertainty and the Evolution of Cooperation. *Journal of Conflict Resolution* 37(4):709–734.

Bendor, J., R. M. Kramer, and S. Stout. 1991. When in Doubt. Cooperation in a Noisy Prisoners-Dilemma. *Journal of Conflict Resolution* 35(4):691–719.

Bendor, J., R. Kramer, and P. Swistak. 1996. Cooperation under Uncertainty: What Is New, What Is True, and What Is Important. *American Sociological Review* 61(2):333–338.

Bendor, J., and D. Mookherjee. 1987. Institutional Structure and the Logic of Ongoing Collective Action. *American Political Science Review* 81(1):129–154.

Bendor, J., and P. Swistak. 1997. The Evolutionary Stability of Cooperation. *American Political Science Review* 91(2):290–307.

——— 2001. The Evolution of Norms. *American Journal of Sociology* 106(2):1493–1545.

Bersaglieri, T., et al. 2004. Genetic Signatures of Strong Recent Positive Selection at the Lactase Gene. *American Journal of Human Genetics* 74(6):1111–1120.

Bliege Bird, R. L., E. A. Smith, and Douglas W. Bird. 2001. The Hunting Handicap: Costly Signaling in Human Foraging Societies. *Behavioral Ecology and Sociobiology* 50:9–19.

Boerlijst, Maarten C., Martin A. Nowak, and Karl Sigmund. 1997. The Logic of Contrition. *Journal of Theoretical Biology* 185:281–293.

Bolton, Gary & Ockenfels, Axel (1999) A Theory of Equity, Reciprocity and Competition. *American Economic Review* 90(1):166–194.

Booth, Heather. 1999. Pacific Island Suicide in Comparative Perspective. *Journal of Biosocial Science* 31:433–448.

Booth, J. 1998. Immigrants Form Family Businesses Out of Necessity. *Los Angeles Business Journal*. http://www.findarticles.com/p/articles/mi_m5072/is_48_20/ai_53449804#-continue. Accessed July 2005.

Boyd, Robert. 1989. Mistakes Allow Evolutionary Stability in the Repeated Prisoner's Dilemma Game. *Journal of Theoretical Biology* 136:47–56.

———— 1992. The Evolution of Reciprocity when Conditions Vary. In *Coalitions in Humans and Other Animals*, ed. F. DeWaal and A. H. Harcourt, pp. 473–492. New York: Oxford University Press.

Boyd, Robert, et al. 2003. The Evolution of Altruistic Punishment. *Proceedings of the National Academy of Sciences* 100(6):3531–3535.

Boyd, Robert, and Jeffrey P. Lorderbaum. 1987. No Pure Strategy Is Evolutionarily Stable in the Repeated Prisoner's Dilemma Game. *Nature* 32:58–59.

Boyd, Robert, and Peter Richerson. 1985. *Culture and the Evolutionary Process.* Chicago: University of Chicago Press.

———— 1988. The Evolution of Reciprocity in Sizable Groups. *Journal of Theoretical Biology* 132:337–356.

———— 1989. The Evolution of Indirect Reciprocity. *Social Networks* 11:213–236.

———— 1992. Punishment Allows the Evolution of Cooperation (or Anything Else) in Sizable Groups. *Ethology and Sociobiology* 13(3):171–195.

———— 2002. Group Beneficial Norms Can Spread Rapidly in a Structured Population. *Journal of Theoretical Biology* 215:287–296.

———— 2005. Solving the Puzzle of Human Cooperation. In *Evolution and Culture*, ed. S. Levinson. Cambridge: MIT Press. pp. 105–132.

Boyer, Pascal. 2001. *Religion Explained: The Evolutionary Origins of Religious Thought.* New York: Basic Books.

Brandt, Hannelore, and Karl Sigmund. 2004. The Logic of Reprobation: Assessment and Action Rules for Indirect Reciprocation. *Journal of Theoretical Biology* 231(4):475–486.

Branswell, H. 2005. *Ontario Reports Large Rubella Outbreak, Warns Could Be Tip of Iceberg.* Children's Health News. Available at: http://mediresource.sympatico.ca/channel_health_news_detail.asp?channel_id=9&menu_item_id=4&news_id=6574. Accessed July 20, 2005.

Brown, Donald E. 1991. *Human Universals.* New York: McGraw Hill.

Bryan, James H. 1971. Model Affect and Children's Imitative Altruism. *Child Development* 42(6):2061–2065.

Bryan, James H., Joel Redfield, and Sandra Mader. 1971. Words and Deeds about Altruism and the Subsequent Reinforcement Power of the Model. *Child Development* 42(5):1501–1508.

Bryan, James H. & Test, Mary Ann. 1967. Models and Helping: Naturalistic Studies in Aiding Behavior. *Journal of Personality and Social Psychology* 6(4):400–407.

Bryan, James H., and Nancy H. Walbek. 1970a. The Impact of Words and Deeds concerning Altruism upon Children. *Child Development* 41(3):747–757.

———— 1970b. Preaching and Practicing Generosity: Children's Actions and Reactions. *Child Development* 41(2):329–353.

Camerer, Colin. 2003. *Behavior Game Theory: Experiments in Strategic Interaction.* Princeton, NJ: Princeton University Press.

———— 1995. Individual Decision Making. In *The Handbook of Experimental Economics*, ed. J. H. Kagel and A. E. Roth, pp. 587–703. Princeton, NJ: Princeton University Press.

Cameron, Lisa. 1999. Raising the Stakes in the Ultimatum Game: Experimental Evidence from Indonesia. *Economic Inquiry* 37(1):47–59.

Carpenter, Jeffrey. 2004. When in Rome: Conformity and the Provision of Public Goods. *Journal of Socio-Economics* 4(4):395–408.

Carpenter, Jeffrey, Stephen Burks, and Eric Verhoogen. 2005. Comparing Students to Workers: The Effects of Social Framing on Behavioral in Distribution Games. In

Field Experiments in Economics, ed. J. Carpenter, G. Harrison, and J. List, pp. 261–190. Greenwich, CT: JAI Press.

Cavalli-Sforza, Luca Luigi. 1973. Some Current Problems of Human Population Genetics. *American Journal of Human Genetics* 25(1):82–104.

CBS News. 2004. *AIDS Fund Faces Shortfall*. Available at: http://www.cbsnews.com/ stories/2004/08/19/health/main636976.shtml. Accessed July 21, 2005.

Chagnon, Napolean, and William Irons. 1979. *Evolutionary Biology and Human Social Behavior: An Anthropological Perspective*. North Scituate, MA: Duxbury Press.

Chaldean Federation of America. 1999. *Chaldean History and Culture*.

Chalmers, D. K., W. C. Horne, and M. E. Rosenbaum. 1963. Social Agreement and the Learning of Matching Behavior. *Journal of Abnormal and Social Psychology* 66:556–561.

Charness, Gary, and Matthew Rabin. 2002. Social Preferences: Some Simple Tests of a New Model. *Quarterly Journal of Economics* 117(3):817–869.

CIA. 2005. The World Factbook. Available at: http://www.cia.gov/cia/publications/ factbook/geos/us.html. Accessed October 31, 2005.

Cohen, John. 2005. If Only It Were This Easy. Available at: http://www.technologyreview .com/articles/05/02/issue/forward_vaccination.asp. Accessed October 31, 2005.

Cooper, Russell, et al. 1996. Cooperation without Reputation: Experimental Evidence from Prisoner's Dilemma Games. *Game and Economic Behavior* 12(12):187–218.

Cosmides, Leda. 1989. The Logic of Social Exchange: Has Natural Selection Shaped How Humans Reason? Studies with the Wason Selection Task. *Cognition* 31(3):187–276.

Coultas, Julie. 2004. When in Rome. An Evolutionary Perspective on Conformity. *Group Processes and Intergroup Relations* 7(4):317–331.

Cronk, Lee, Napolean Chagnon, and William Irons, eds. 2000. *Adaptation and Human Behavior*. New York: Aldine De Gruyter.

Cummings, Ronald, and Paul Ferraro. 2002. Cultural Diversity, Discrimination and Economic Outcomes: An Experimental Analysis. Available at: http://epp.gsu.edu/ pferraro/docs/FerraroCummingsEJFinalSubmit.pdf.

Daly, Martin, and Margo Wilson. 1999. *The Truth about Cinderella: A Darwinian View of Parental Love*. New Haven, CT: Yale University Press.

Darwin, Charles. 1981. *The Descent of Man in Relation to Sex*. Princeton, NJ: Princeton University Press.

Dawes, R. 1991. Social Dilemmas, Economic Self-interest, and Evolutionary Theory. In *Frontiers of Mathematical Psychology: Essays in Honor of Clyde Coombs*, ed. Donald Brown and *J. E. Keith Smith*, pp. 53–79. New York: Springer-Verlag.

Dawkins, Richard. 1976. *The Selfish Gene*. Oxford: Oxford University Press.

DeBruine, Lisa. 2002. Facial Resemblance Enhances Trust. *Proceedings of the Royal Society of London Series B-Biological Sciences* 269:1307–1312.

Denant-Boemont, Laurent, David Masclet, and Charles Noussair. 2005. Anonymity in Punishment, Revenge and Cooperation: A Public Goods Experiment. Available at: http://userwww.service.emory.edu/~cnoussa/Punishment%20August%202005.pdf.

Diamond, Jared M. 1997. *Guns, Germs, and Steel: The Fates of Human Societies*. New York: W. W. Norton.

Durham, William H. 1991. *Coevolution: Genes, Culture, and Human Diversity*. Stanford, CA: Stanford University Press.

Durkheim, Emile. 1933. *The Division of Labor in Society*. New York: Free Press.

Eckel, C., and P. Grossman. 1996. Altruism in Anonymous Dictator Games. *Games and Economic Behavior* 16:181–191.

Edgerton, Robert B. 1992. *Sick Societies: Challenging the Myth of Primitive Harmony*. New York: Free Press.

Elliot, R., and R. Vasta. 1970. The Modeling of Sharing: Effects Associated with Vicarious Reinforcement, Symbolization, Age, and Generalization. *Journal of Experimental Child Psychology* 10:8–15.

Engelmann, Dirk, and Urs Fischbacher. 2002. Indirect Reciprocity and Strategic Reputation Building in an Experimental Helping Game. University of Zurich. Available at: http://www.iew.unizh.ch/wp/iewwp132.pdf.

The Environmental Literacy Council. 2005. *Kyoto Protocol*. Available at: http://www.enviroliteracy.org/article.php/278.html Updated July 11, 2005. Accessed August 18, 2005.

Farrell, Joseph, and Roger Ware. 1989. Evolutionary Stability in the Repeated Prisoner's Dilemma. *Theoretical Population Biology* 36:161–166.

Fehr, Ernst, and Urs Fischbacher. 2003. The Nature of Human Altruism. *Nature* 425:785–791.

———— 2004. Third Party Sanction and Social Norms. *Evolution and Human Behavior* 25(2004):63–87.

Fehr, E., and S. Gächter. 2000. Cooperation and Punishment in Public Goods Experiments. *American Economic Review* 90(4):980–995.

———— 2002. Altruistic Punishment in Humans. *Nature* 415:137–140.

———— 2005. Egalitarian Motive and Altruistic Punishment (Reply). *Nature* 433:E1–E2.

Fehr, Ernst, and Joseph Henrich. 2003. Is Strong Reciprocity a Maladaptation? In *Genetic and Cultural Evolution of Cooperation*, ed. P. Hammerstein. Cambridge: MIT Press.

Fehr, E., and K. Schmidt. 1999. A Theory of Fairness, Competition, and Cooperation. *Quarterly Journal of Economics* 114:817–868.

Fehr, Ernst, et al. 2002. A Nation-Wide Laboratory: Examining Trust and Trustworthiness by integrating behavioral experiments into representative surveys. Available at: http://www.iew.unizh.ch/wp/iewwp141.pdf.

Fessler, Daniel M. T., and C. D. Navarrete. 2004. Third-Party Attitudes toward Incest: Evidence for the Westermarck Effect. *Evolution and Human Behavior* 25(5):277–294.

Fiske, Alan. 1991. *Structures of Social Life*. New York: Free Press.

———— 1992. The Four Elementary Forms of Sociality: Framework for a Unified Theory of Social Relations. *Psychological Review* 99(4):689–723.

Flannery, K. V., and Joyce Marcus. 2000. Formative Mexican Chiefdoms and the Myth of the "Mother Culture." *Journal of Anthropological Archaeology* 19:1–37.

Forsythe, Robert, et al. 1994. Fairness in Simple Bargaining Experiments. *Games and Economic Behavior* 6:347–369.

Frank, Steven. 1995. George Price's Contributions to Evolutionary Genetics. *Journal of Theoretical Biology* 175:373–388.

———— 1997. The Price Equation, Fisher's Fundamental Theorem, Kin Selection, and Causal Analysis. *Society for the Study of Evolution* 51(6):1712–1729.

———— 1998. *Foundations of Social Evolution*. Princeton, NJ: Princeton University Press.

Gächter, Simon, Herrmann Benedikt, and Christian Thoni. 2005. Cross-Cultural Differences in Norm Enforcement. *Behavioral and Brain Sciences*, 85:822–823.

Gaulin, Steven J. C., Donald H. McBurney, and Stephanie L. Brakeman-Wartell. 1997. Matrilateral Biases in the Investment of Aunts and Uncles: A Consequence and Measure of Paternity Uncertainty. *Human Nature* 8(2):139–151.

Gigerenzer, Gerd, and Klaus Hug. 1992. Domain-Specific Reasoning: Social Contracts, Cheating, and Perspective Change. *Cognition* 43(2):127–171.

Gil-White, Francisco. 1999. How Thick Is Blood? The Plot Thickens: If Ethnic Actors Are Primordialists, What Remains of the Circumstantialist/Primordialist Controversy? *Ethnic and Racial Studies* 22(5):789–820.

———— 2001. Are Ethnic Groups Biological "Species" to the Human Brain? Essentialism in our Cognition of Some Social Categories. *Cultural Anthropology* 42(4):515–554.

———— 2004. Ultimatum Game with an Ethnicity Manipulation: Results from Khovdiin Bulgan Cum, Mongolia. In *Foundations of Human Sociality: Economic Experiments and Ethnographic Evidence from Fifteen Small-Scale Societies*, ed. J. Henrich, R. Boyd, S. Bowles, et al., pp. 260–304. New York: Oxford University Press.

Gintis, H. 2003a. Altruism and Emotions. *Behavioral and Brain Sciences* 25(2):258.

———— 2003b. The Hitchhiker's Guide to Altruism: Gene-Culture Coevolution, and the Internalization of Norms. *Journal of Theoretical Biology* 220(4):407–418.

Gintis, Herbert, et al. 2003. Explaining Altruistic Behavior in Humans. *Evolution and Human Behavior* 24:153–172.

Gowdy, John, Raluca Iorgulescu, and Stephen Onyeiwu. 2003. Fairness and Retaliation in a Rural Nigerian Village. *Journal of Economic Behavior and Organization* 52: 469–479.

Grafen, Alan. 1985. Hamilton's Rule OK. *Nature* 318:310–311.

Greenfield, Norman, and James T. Kuznicki. 1975. Implied Competence, Task Complexity, and Imitative Behavior. *Journal of Social Psychology* 95:251–261.

Grusec, Joan E. 1971. Power and the Internalization of Self-Denial. *Child Development* 42(1):93–105.

Grusec, Joan E., Peter Saas-Kortsaak, and Zita M. Simutis. 1978. The Role of Example and Moral Exhortation in the Training of Altruism. *Child Development* 49(3): 920–923.

Gurven, Michael. 2001. Reservation Food Sharing among the Ache of Paraguay. *Human Nature* 12(4):273–297.

———— 2004. Does Market Exposure Affect Economic Game Behavior? The Ultimatum Game and the Public Goods Game among the Tsimane' of Bolivia. In *Foundations of Human Sociality: Economic Experiments and Ethnographic Evidence from Fifteen Small-Scale Societies*, ed. J. Henrich, R. Boyd, S. Bowles, et al., pp. 194–231. New York: Oxford University Press.

Gurven, Michael, Kim Hill, and Hillard Kaplan. 2002. From Forest to Reservation: Transitions in Food Sharing Behavior among the Ache of Paraguay. *Journal of Anthropological Research* 58(1):93–120.

Haiman, F. S. 1949. An Experimental Study of the Effects of Ethos on Public Speaking. *Speech Monographs* 16:190–202.

Hamilton, William D. 1964. The Genetical Evolution of Social Behaviour. *Journal of Theoretical Biology* 7:1–16.

———— 1972. Altruism and Related Phenomena, Mainly in the Social Insects. *Annual Review of Ecology and Systematics* 3:192–232.

———— 1975. Innate Social Aptitudes of Man: An Approach from Evolutionary Genetics. In *Biosocial Anthropology*, ed. R. Fox, pp. 133–156. London: Malaby Press.

Hammerstein, Peter. 2003. Why Is Reciprocity So Rare in Social Animals? In *Genetic and Cultural Evolution of Cooperation*, ed. P. Hammerstein, pp. 83–94. Cambridge: MIT Press.

Harbaugh, William T., and Kate Krause. 2000. Children's Altruism in Public Goods and Dictator Experiments. *Economic Inquiry* 38(1):95–109.

Harbaugh, William T., Kate Krause, and Steven G. Liday. 2002. Bargaining by Children. Available at: http://economics.uoregon.edu/papers/UO-2002-4_Harbaugh_Kid_Bargaining.pdf.

Harris, Judith Rich. 1995. Where Is the Child's Environment? A Group Socialization Theory of Development. *Psychological Review* 102:458–489.

——— 1998. *The Nurture Assumption: Why Children Turn Out the Way They Do*. New York: Touchstone.

Harris, Mary B. 1970. Reciprocity and Generosity: Some Determinants of Sharing in Children. *Child Development* 41:313–328.

——— 1971. Models, Norms, and Sharing. *Psychological Reports* 29:147–153.

Hartshorne, Hugh, and Mark May. 1928a. *Studies in Deceit*. Vol. 1 of *Studies in the Nature of Character*. New York: Macmillan.

——— 1928b. *Studies in Self-Control*. Vol. 2 of *Studies in the Nature of Character*. New York: Macmillan.

Hartshorne, Hugh, Mark May, and F. Shuttleworth. 1930. *Studies in the Organization of Character*. Vol. 3 of *Studies in the Nature of Character*. New York: Macmillan.

Hartup, W. W., and B. Coates. 1967. Imitation of Peers as a Function of Reinforcement from the Peer Group and Rewardingness of the Model. *Child Development* 1967(38):1003–1016.

Hawkes, Kristen, James F. O'Connell, and Nicholas Burton Jones. 2001. Hadza Meat Sharing. *Evolution and Human Behavior* 22:113–142.

Henrich, Joseph. 2000. Does Culture Matter in Economic Behavior: Ultimatum Game Bargaining among the Machiguenga. *American Economic Review* 90(4):973–980.

——— 2002. Decision-Making, Cultural Transmission, and Adaptation in Economic Anthropology. In *Theory in Economic Anthropology*, ed. J. Ensminger, pp. 251–295. Lanham, MD: Rowman and Littlefield.

——— 2004a. Cultural Group Selection, Coevolutionary Processes, and Large-Scale Cooperation. *Journal of Economic Behavior and Organization* 53:3–35.

——— 2004b. Demography and Cultural Evolution: Why Adaptive Cultural Processes Produced Maladaptive Losses in Tasmania. *American Antiquity* 69(2):197–214.

——— 2006. Cooperation, Punishment, and the Evolution of Human Institutions. *Science* 312:60–61.

——— Forthcoming. A Cultural Species. In *Explaining Culture Scientifically*, ed. M. Brown. Seattle: University of Washington Press.

Henrich, Joseph, Richard McElreath, Abigail Barr, Jean Ensminger, Clark Barrett, Alexander Bolyanatz, Juan Camilo Cardenas, Michael Gurven, Edwin Gwako, Natalie Henrich, Carolyn Lesorogol, Frank Marlowe, David Tracer, and John Ziker. 2006. Costly Punishment across Human Societies. *Science* 312:1767–1770.

Henrich, Joseph, Samuel Bowles, Robert Boyd, Astrid Hopfensitz, Peter Richerson, Karl Sigmund, Eric A. Smith, Franz Weissing, and H. Peyton Young. 2003. The Cultural and Genetic Evolution of Human Cooperation. In *Genetic and Cultural Evolution of Cooperation*, ed. P. Hammerstein, pp. 445–468. Cambridge: MIT Press.

Henrich, Joseph, and Robert Boyd. 1998. The Evolution of Conformist Transmission and the Emergence of Between-group Differences. *Evolution and Human Behavior* 19:215–242.

——— 2001. Why People Punish Defectors: Weak Conformist Transmission Can Stabilize Costly Enforcement of Norms in Cooperative Dilemmas. *Journal of Theoretical Biology* 208:79–89.

——— 2002. On Modeling Cultural Evolution: Why Replicators Are Not Necessary for Cultural Evolution. *Journal of Cognition and Culture* 2(2):87–112.

Henrich, Joseph, R. Boyd, C. Bowles, et al., eds. 2004. *Foundations of Human Sociality: Economic Experiments and Ethnographic Evidence from Fifteen Small-Scale Societies*. Oxford: Oxford University Press.

Henrich, Joseph, Robert Boyd, Samuel Bowles, Colin Camerer, Ernst Fehr, et al. 2005. "Economic Man" in Cross-cultural Perspective: Behavioral Experiments in 15 Small-Scale Societies. *Behavioral and Brain Sciences* 28:796–855.

Henrich, Joseph, and Francisco Gil-White. 2001. The Evolution of Prestige: Freely Conferred Deference as a Mechanism for Enhancing the Benefits of Cultural Transmission. *Evolution and Human Behavior* 22(3):165–196.

Henrich, Joseph, and Richard McElreath. 2003. The Evolution of Cultural Evolution. *Evolutionary Anthropology* 12(3):123–135.

Henrich, Joseph, and Natalie Smith. 2004. Comparative Experimental Evidence from Machiguenga, Mapuche, and American Populations. In *Foundations of Human Sociality: Economic Experiments and Ethnographic Evidence from Fifteen Small-Scale Societies*, ed. J. Henrich, R. Boyd, S. Bowles, et al., pp. 125–167. Oxford: Oxford University Press.

Herbert, Emily W., Donna M. Gelfand, and Donald Hartmann. 1969. Imitation and Self-Esteem as Determinants of Self-Critical Behavior. *Child Development* 40:421–430.

Hill, Kim. 2002. Altruistic Cooperation during Foraging by the Ache and the Evolved Human Predisposition to Cooperate. *Human Nature* 13(1):105–128.

Hirshleifer, Jack, and Juan Carlos Martinez. 1988. What Strategies Can Support the Evolutionary Emergence of Cooperation? *Journal of Conflict Resolution* 32(2): 367–398.

Hoffman, Elizabeth, et al. 1994. Preferences, Property Rights, and Anonymity in Bargaining Games. *Game and Economic Behavior* 7:346–380.

Holden, C., and R. Mace. 1997. Phylogenetic Analysis of the Evolution of Lactose Digestion in Adults. *Human Biology* 69(5):605–628.

Hruschka, Daniel, and Joseph Henrich. 2006. Friendship, Cliquishness, and the Emergence of Cooperation. *Journal of Theoretical Biology* 239(1):1–15.

Hua, Vanessa. 2002. Immigrants Find Family Business Can Fuel Stereotypes–Early Arrivals Were Forced to Find Niche Work. *San Francisco Chronicle*. http://www.sfgate.com/cgi-bin/article.cgi?file=/chronicle/archive/2002/12/31/MN106222.DTL. Accessed July 2005.

Huck, Steffen, and Jorg Oeschssler. 1999. The Indirect Evolutionary Approach to Explaining Fair Allocations. *Games and Economic Behavior* 28:13–24.

Jammo, Sarhad. 2001. Presentation at Orchard Lake Middle School. Orchard Lake: Michigan.

Johnson, Allen. 2003. *Families of the Forest: Matsigenka Indians of the Peruvian Amazon*. Los Angeles: University of California.

Johnson, Allen, and Timothy Earle. 2000. *The Evolution of Human Societies*. Stanford, CA: Stanford University Press.

Johnson, J. D., et al. 1977. Lactose-Malabsorption among Pima Indians of Arizona. *Gastroenterology* 73(6):1299–1304.

Jonas, Klaus. 1992. Modelling and Suicide: A Test of the Werther Effect. *British Journal of Social Psychology* 31:295–306.

Joshi, N V. 1987. Evolution of Cooperation by Reciprocation within Structured Demes. *Journal of Genetics* 66(1):69–84.

Kameda, Tatsuya, and D. Nakanishi. 2002. Cost-Benefit Analysis of Social/Cultural Learning in a Non-Stationary Uncertain Environment: An Evolutionary Simulation and an Experiment with Human Subjects. *Evolution and Human Behavior* 23:373–393.

Kamoo, Ray. 1999. *Ancient and Modern Chaldean History: A Comprehensive Bibliography of Sources*. Lanham, MD: Scarecrow Press.

Keller, Laurent, and Kenneth G. Ross. 1998. Selfish Genes, a Green Beard in the Red Fire Ant. *Nature* 394:573–575.

Kellogg, Wintrop, and Luella Kellogg. 1933. *The Ape and the Child: A Study of Environmental Influence upon Early Behavior*. New York: McGraw-Hill.

Kelly, Raymond C. 1985. *The Nuer Conquest*. Ann Arbor: University of Michigan Press.

Kelman, H. C. 1958. Compliance, Identification, and Internalization: Three Processes of Attitude Change. *Journal of Conflict Resolution* 2:51–60.

Kessler, Ronald C., G. Downey, and Horst Stipp. 1988. Clustering of Teenage Suicides after Television News Stories about Suicide: A Reconsideration. *American Journal of Psychiatry* 145:1379–1383.

Kroll, Yoram, and Haim Levy. 1992. Further Tests of the Separation Theorem and the Capital Asset Pricing Model. *American Economic Review* 82(3):664–670.

Labov, William. 1972. *Sociolinguistic Patterns*. Philadelphia: University of Pennsylvania Press.

——— 1980. The Social Origins of Sound Change. In *Locating Language in Time and Space*, ed. W. Labov. New York: Academic Press.

Laland, Kevin, John Odling-Smee, and Marcus Feldman. 2000. Niche Construction, Biological Evolution, and Cultural Change. *Behavioral and Brain Sciences* 23(1):131–175.

Ledyard, John O. 1995. Public Goods: A Survey of Experimental Research. In *The Handbook of Experimental Economics*, pp. 111–194. Princeton, NJ: Princeton University Press.

Leimar, Olof, and Peter Hammerstein. 2001. Evolution of Cooperation through Indirect Reciprocity. Proceedings. *Biological Sciences* 268(1468):745–753.

Levinson, Stephen C. 2003. *Space in Language and Cognition*. Cambridge: Cambridge University Press.

Linn, Virginia. 2005. Once Nearly Eradicated, Polio Makes a Comeback in Africa. In Post-gazette.com. April 3, 2005. Available at: http://www.post-gazette.com/pg/05093/481043.stm. Accessed October 31, 2005.

Lorberbaum, Jeffrey. 1994. No Strategy Is Evolutionarily Stable in the Repeated Prisoner's Dilemma. *Journal of Theoretical Biology* 168:117–130.

Lorberbaum, Jeffrey P., et al. 2002. Are There Really No Evolutionary Stable Strategies in the Iterated Prisoner's Dilemma? *Journal of Theoretical Biology* 214:155–169.

Manson, Joseph H., and Richard Wrangham. 1991. Intergroup Aggression in Chimpanzees and Humans. *Current Anthropology* 32:369–390.

Marland, G., T. A. Boden, and R. J. Andres. 2005. Global, Regional, and National CO_2 Emissions. In *Trends: A Compendium of Data on Global Change*. Carbon Dioxide Information Analysis Center, Oak Ridge National Laboratory, U.S. Department of Energy, Oak Ridge, TN. Available at: http://cdiac.esd.ornl.gov/trends/emis/tre_usa.htm.

Masclet, David, et al. 2003. Monetary and Nonmonetary Punishment in the Voluntary Contributions Mechanism. *American Economic Review* 93(1):366–380.

Mausner, B. 1954. The Effect of Prior Reinforcement on the Interaction of Observed Pairs. *Journal of Abnormal and Social Psychology* 49:65–68.

Mausner, B., and B. L. Bloch. 1957. A Study of the Additivity of Variables Affecting Social Interaction. *Journal of Abnormal and Social Psychology* 54:250–256.

Maynard Smith, John. 1982. *Evolution and the Theory of Games*. Cambridge: Cambridge University Press.

McElreath, Richard, Robert Boyd, and Peter Richerson. 2003. Shared Norms Can Lead to the Evolution of Ethnic Markers. *Current Anthropology* 44(1):123–129.

McElreath, Richard, Mark Lubell, Peter Richerson, Timothy Waring, William Baum, Edward Edsten, Charles Efferson, and Brian Paciotti. 2005. Applying Evolutionary Models to the Laboratory Study of Social Learning. *Evolution and Human Behavior* 26:483–508.

McKelvey, R., and T. Page. 1990. Public and Private Information: An Experimental Study of Information Pooling. *Econometrica* 58(6):1321–1339.

Meltzoff, Andrew N., and Wolfgang Prinz, eds. 2002. *The Imitative Mind*. New York: Cambridge University Press.

Midlarsky, Elizabeth, and James H. Bryan. 1972. Affect Expressions and Children's Imitative Altruism. *Journal of Experimental Child Psychology* 6:195–203.

Midlarsky, Elizabeth, James H. Bryan, and Philip Brickman. 1973. Aversive Approval: Interactive Effects of Modeling and Reinforcement on Altruistic Behavior. *Child Development* 44(2):321–328.

Miller, N. E., and J. Dollard. 1941. *Social Learning and Imitation*. New Haven, CT: Yale University Press.

Mischel, W., and R. M. Liebert. 1966. Effects of Discrepancies between Observed and Imposed Reward Criteria on their Acquisition and Transmission. *Journal of Personality and Social Psychology* 3:45–53.

Mulcare, C. A., et al. 2004. The T Allele of a Single-Nucleotide Polymorphism 13.9 kb Upstream of the Lactase Gene (LCT) (C-13.9kbT) Does Not Predict or Cause the Lactase-Persistence Phenotype in Africans. *American Journal of Human Genetics* 74(6):1102–1110.

Mullen, B., C. Cooper, and J. E. Driskell. 1990. Jaywalking as a Function of Model Behavior. *Personality and Social Psychology Bulletin* 16: 320–330.

Nettle, Daniel, and Robin I. M. Dunbar. 1997. Social Markers and the Evolution of Reciprocal Exchange. *Current Anthropology* 38(1):93–99.

Nisbett, Richard E. 2003. *The Geography of Thought: How Asians and Westerners Think Differently and Why*. New York: Free Press.

Nowak, Martin A., and Karl Sigmund. 1998a. The Dynamics of Indirect Reciprocity. *Journal of Theoretical Biology* 194:561–574.

——— 1998b. Evolution of Indirect Reciprocity by Image Scoring. *Nature* 393:573–577.

Offerman, Theo, Jan Potters, and Joep Sonnemans. 2002. Imitation and Belief Learning in an Oligopoly Experiment. *Review of Economic Studies* 69(4):973–998.

Offerman, Theo, and Joep Sonnemans. 1998. Learning by Experience and Learning by Imitating Others. *Journal of Economic Behavior and Organization* 34(4):559–575.

Ohtsuki, Hisashi, and Yoh Iwasa. 2004. How Should We Define Goodness? Reputation Dynamics in Indirect Reciprocity. *Journal of Theoretical Biology* 231(1):107–120.

Oota, H., et al. 2001. Human mtDNA and Y-chromosome Variation Is Correlated with Matrilocal versus Patrilocal Residence. *Nature Genetics* 29(1):20–21.

Ostrom, E., R. Gardner, and J. Walker. 1994. *Rules, Games, and Common-Pool Resource Problems*. Ann Arbor: University of Michigan Press.

Panchanathan, Karthic, and Robert Boyd. 2003. A Tale of Two Defectors: The Importance of Standing for the Evolution of Indirect Reciprocity. *Journal of Theoretical Biology* 224:115–126.

——— 2004. Indirect Reciprocity Can Stabilize Cooperation without the Second-Order Free Rider Problem. *Nature* 432:499–502.

Park, Kyeyoung.1997. *The Korean American Dream: Immigrants and Small Business in New York City*. Ithaca: Cornell University Press.

Parr, Lisa, and Frans de Waal. 1999. Visual Kin Recognition in Chimpanzees. *Nature* 399:647–648.

Pingle, M. 1995. Imitation vs. Rationality: An Experimental Perspective on Decision-Making. *Journal of Socio-Economics* 24:281–315.

Platek, Steven M., et al. 2002. Reactions to Children's Faces: Resemblance Affects Males More Than Females. *Evolution and Human Behavior* 23:159–166.

———— 2003. How Much Paternal Resemblance Is Enough? Sex Differences in Hypothetical Investment Decisions but Not in the Detection of Resemblance. *Evolution and Human Behavior* 24:81–87.

Plomin, Robert, John Defries, and Gerald E. McLearn. 2001. *Behavioral Genetics*. New York: W. H. Freeman.

Presbie, Robert J., and Paul F. Coiteux. 1971. Learning to Be Generous or Stingy: Imitation of Sharing Behavior as a Function of Model Generosity and Vicarious Reinforcement. *Child Development* 42(4):1033–1038.

Price, George. 1970. Selection and Covariance. *Nature* 227:520–521.

———— 1972. Extensions of Covariance Selection Mathematics. *Annals of Human Genetics* 35:485–490.

Price, Michael, John Tooby, and Leda Cosmides. 2002. Punitive Sentiment as an Anti–Free Rider Psychological Device. *Evolution and Human Behavior* 23:203–231.

Public Health Agency of Canada. 2004. *Thimerosal in Vaccines and Autism*. Available at: https://www.phac-aspc.gc.ca/dird-dimr/q_a_thimerosal_e.html. Accessed July 19, 2005.

Public Health News Center. 2005. *Why Are Parents Refusing to Vaccinate School-Aged Children?* Johns Hopkins Bloomberg School of Public Health. Available at: http://www.jhsph.edu/publichealthnews/articles/2005/salmon_vaccines.html. Accessed July 19, 2005.

Putnam, Robert. 2000. *Bowling Alone: The Collapse and Revival of American Community*. New York: Simon and Schuster.

Queller, David C. 1992. A General Model For Kin Selection. *Evolution* 46(2):376–380.

Rice, M. E., and Joan E. Grusec. 1975. Saying and Doing: Effects of Observer Performance. *Journal of Personality and Social Psychology* 32:584–593.

Richerson, Peter, and Robert Boyd. 1998. The Evolution of Ultrasociality. In *Indoctrinability, Ideology, and Warfare*, ed. I. Eibl-Eibesfeldt and F. K. Salter, pp. 71–96. New York: Berghahn Books.

———— 2000. Complex Societies: The Evolutionary Dynamics of a Crude Superorganism. *Human Nature* 10:253–289.

———— 2005. *Not by Genes Alone: How Culture Transformed Human Evolution*. Chicago: University of Chicago Press.

Richerson, Peter, Robert Boyd, and Robert L. Bettinger. 2001. Agriculture Impossible During the Pleistocene But Mandatory During the Holocene? A Climate Change Hypothesis. *American Antiquity* 66:387–411.

Richerson, Peter, Robert Boyd, and Joseph Henrich. 2003. The Cultural Evolution of Cooperation. In *The Genetic and Cultural Evolution of Cooperation*, ed. P. Hammerstein, pp. 357–388. Cambridge: MIT Press.

Ritchie, Elaine, and Jerry E. Phares. 1969. Attitude Change as a Function of Internal External Control and Communicator Status. *Journal of Personality* 37(3):429–443.

Rogers, Alan. 1989. Does Biology Constrain Culture? *American Anthropologist* 90:819–831.

Rogers, Everett M. 1995. *Diffusion of Innovations*. New York: Free Press.

Rosenbaum, M., and R. R. Blake. 1955. The Effect of Stimulus and Background Factors on the Volunteering Response. *Journal of Abnormal and Social Psychology* 50:193–196.

Rosenbaum, Milton E., and Irving F. Tucker. 1962. The Competence of the Model and the Learning of Imitation and Nonimitation. *Journal of Experimental Psychology* 63(2):183–190.

Rosenhan, David, and Glenn M. White. 1967. Observation and Rehearsal as Determinants of Prosocial Behavior. *Journal of Personality and Social Psychology* 5(4): 424–431.

Rosenthal, F. 1978. Aramaic Studies During the Past Thirty Years. *Journal of Near Eastern Studies* 37:81–82.

Ross, L., and R. E. Nisbett. 1991. *The Person and the Situation: Perspective of Social Psychology*. Philadelphia: Temple University Press.

Roth, Alvin E., et al. 1991. Bargaining and Market Behavior in Jerusalem, Ljubljana, Pittsburgh, and Tokyo: An Experimental Study. *American Economic Review* 81(5):1068–1095.

Rubinstein, Donald H. 1983. Epidemic Suicide among Micronesian Adolescents. *Social Science Medicine* 17(10):657–665.

Rushton, J. P. 1975. Generosity in Children: Immediate and Long-Term Effects of Modeling, Preaching, and Moral Judgment. *Journal of Personality and Social Psychology* 31:459–466.

Rushton, J. P., and A. C. Campbell. 1977. Modeling, Vicarious Reinforcement, and Extraversion on Blood Donating in Adults. Immediate and Long-Term Effects. *European Journal of Social Psychology* 7:297–306.

Ryckman, R. M., W. C. Rodda, and M. F. Sherman. 1972. Locus of Control and Expertise Relevance as Determinants of Changes in Opinion about Student Activisim. *Journal of Social Psychology* 88:107–114.

Sahlins, Marshall. 1977. *The Use and Abuse of Biology*. Ann Arbor: University of Michigan Press.

Sanders, J., and V. Nee. 1996. Immigrant Self-Employment: The Family as Social Capital and the Value of Human Capital. *American Sociological Review*, 61:231–249.

Sarafa, Josephine Saroki. n.d. The Chaldean Americans: Past and Present: St. Thomas the Apostle. Chaldean Catholic Diocese of America.

Schachter, S., and R. Hall. 1952. Group-Derived Restraints and Audience Persuasion. *Human Relations* 5:397–406.

Schlag, Karl H. 1998. Why Imitate, and If So, How? A Boundedly Rational Approach to Multi-Armed Bandits. *Journal of Economic Theory* 78:130–156.

———— 1999. Which One Should I Imitate? *Journal of Mathematical Economics* 31(4):493–527.

Scrimshaw, N., and E. Murray. 1988. The Acceptability of Milk and Milk Products in Populations with a High Prevalence of Lactose Intolerance. *American Journal of Clinical Nutrition* 48:1079–1159.

Seinen, Ingrid, and Arthur Schram. 2006. Social Status and Group Norms: Indirect Reciprocity in a Helping Experiment. *European Economic Review* 3:581–602.

Sengstock, Mary Cay. 1999. *Chaldean Americans: Changing Conceptions of Ethnic Identity*. New York: Center for Migration Studies.

Shennan, Stephen. 2003. *Genes, Memes, and Human History: Darwinian Archaeology and Cultural Evolution*. London: Thames and Hudson.

Shepher, J. 1983. *Incest, the Biosocial View*. New York: Academic Press.

Silk, Joan. 1980. Adoption and Kinship in Oceania. *American Anthropologist* 82:799–820.

Simoons, F. J. 1970. Primary Adult Lactose Intolerance and Milking Habit—a Problem in Biologic and Cultural Interrelations. II. A Culture Historical Hypothesis. *American Journal of Digestive Diseases* 15(8):695–710.

Skinner, Jonathan, and Joel Slemrod. 1985. An Economic Perspective on Tax Evasion. *National Tax Journal* 38:345–353.

Smith, Adam. 2000 (1759). *The Theory of Moral Sentiments*. New York: Prometheus Books.

Smith, Eric Alden. 2003. Human Cooperation: Perspectives from Behavioral Ecology. In *Genetic and Cultural Evolution of Cooperation*, ed. P. Hammerstein. Cambridge: MIT Press.

Smith, Eric Alden, and Rebecca Lynn Bliege Bird. 2000. Turtle Hunting and Tombstone Opening: Public Generosity as Costly Signaling. *Evolution and Human Behavior* 21:245–261.

Smith, Jeffrey M, and Paul A. Bell. 1994. Conformity as a Determinant of Behavior in a Resource Dilemma. *Journal of Social Psychology* 134(2):191–200.

Smith, Natalie. 2001. Are Indigenous People Conservationists? Preliminary Results from the Machiguenga of the Peruvian Amazon. *Rationality and Society*. 13(4):429–461.

Soltis, J., R. Boyd, and P. J. Richerson. 1995. Can Group-Functional Behaviors Evolve by Cultural Group Selection? An Empirical Test. *Current Anthropology* 36(3):473–494.

Sowell, Thomas. 1998. *Conquests and Cultures: An International History*. New York: Basic Books.

Stack, Steven. 1987. Celebrities and Suicide: A Taxonomy and Analysis, 1948–1983. *American Sociological Review* 52(3):401–412.

——— 1990. Divorce, Suicide, and the Mass Media: An Analysis of Differential Identification, 1948–1980. *Journal of Marriage and the Family* 52(2):553–560.

——— 1991. Social Correlates of Suicide by Age: Media Impacts. In *Life Span Perspectives of Suicide: Timelines in the Suicide Process*, ed. A. Leenaars, pp. 187–213. New York: Plenum Press.

——— 1996. The Effect of the Media on Suicide: Evidence from Japan, 1955–1985. *Suicide and Life-Threatening Behavior* 26(2):132–142.

——— 2000. Media Impacts on Suicide: A Quantitative Review of 293 Findings. *Social Science Quarterly* 81(4):957–971.

Stark, Rodney. 1997. *The Rise of Christianity: How the Obscure, Marginal Jesus Movement Became the Dominant Religious Force in the Western World in a Few Centuries*. New York: Harper Collins.

Stone, V. E., et al. 2002. Selective Impairment of Reasoning about Social Exchange in a Patient with Bilateral Limbic System Damage. *Proceedings of the National Academy of Sciences of the United States of America* 99(17):11531–11536.

Sugden, Robert. 1986. *The Economics of Rights, Co-operation, and Welfare*. Oxford: Basil and Blackwell.

Sugiyama, L. S., J. Tooby, and L. Cosmides. 2002. Cross-Cultural Evidence of Cognitive Adaptations for Social Exchange among the Shiwiar of Ecuadorian Amazonia. *Proceedings of the National Academy of Sciences* 99(17):11537–11542.

Tannenbaum, P. H. 1956. Initial Attitude toward Source and Concept as Factors in Attitude Change through Communication. *Public Opinion Quarterly* 20:413–425.

The Global Fund. 2005. Available at: http://www.theglobalfund.org/en/. Accessed July 21, 2005.

Thornhill, Randolph, et al. 2003. Major Histocompatibility Complex Genes, Symmetry, and Body Scent Attractiveness in Men and Women. *Behavioral Ecology* 15(5): 668–678.

Tracer, David. 2003. Selfishness and Fairness in Economic and Evolutionary Perspective: An Experimental Economic Study in Papua New Guinea. *Current Anthropology* 44(3):432–438.

Tomasello, Michael. 1999. The Human Adaptation for Culture. *Annual Review of Anthropology*. 28:509–29.

———— 2000. *The Cultural Origins of Human Cognition*. Cambridge: Harvard University Press.

Tooby, John, and Leda Cosmides. 1989. Evolutionary Psychology and the Generation of Culture: Theoretical Considerations. *Ethology and Sociobiology* 10:29–49.

———— 1992. The Psychological Foundations of Culture. In *The Adapted Mind: Evolutionary Psychology and the Generation of Culture*, ed. J. Barkow, L. Cosmides, and J. Tooby, pp. 19–136. New York: Oxford University Press.

Trivers, Robert L. 1971. The Evolution of Reciprocal Alturism. *Quarterly Review of Biology* 46:34–57.

———— 1985. *Social Evolution*. Menlo Park, CA: Benjamin-Cummings.

U.S. Mission to the United Nations in Geneva. 2004. *U.S. Challenges Other Donors to Support Global Disease Fund–AIDS Ambassador Says U.S. Contribution Will Rise If Other Donations Rise*. Press Release, August 19, 2004. Available at: http://www.us-mission.ch/press2004/0819USglobalAIDS.htm. Accessed July 21, 2005.

Wasserman, Ira M., Steven Stack, and Jimmie L. Reeves. 1994. Suicide and the Media: The New York Times's Presentation of Front-Page Suicide Stories between 1910 and 1920. *Journal of Communication* 44(2):64–83.

Wedekind, Claus, and Manfred Milinski. 2000. Cooperation through Image Scoring in Humans. *Science* 288:850–852.

Wilson, David Sloan. 2002. *Darwin's Cathedral: Evolution, Religion, and the Nature of Society*. Chicago: University of Chicago.

Wit, Jorgen. 1999. Social Learning in a Common Interest Voting Game. *Games and Economic Behavior* 26:131–156.

Wolf, Arthur P. 1995. *Sexual Attraction and Childhood Association: A Chinese Brief for Edward Westermarck*. Stanford, CA: Stanford University Press.

World Health Organization. 2005. Massive International Effort Stops Polio Epidemic Across 10 West and Central African Countries. Available at: http://www.who.int/mediacentre/news/releases/2005/pr60/en/index.html. Accessed October 24, 2006.

World Resources Institute. 2005. Earth Trends. The Environmental Information Portal. Available at: http://earthtrends.wri.org/searchable_db/index.cfm?theme=3&variable_ID=470&action=select_countries. Accessed December 12, 2005.

Wu, Jianzhong, and Robert Axelrod. 1995. How to Cope with Noise in the Iterated Prisoner's Dilemma. *Journal of Conflict Resolution* 39(1):183–189.

Yamagishi, Toshio. 1986. The Provision of a Sanctioning System as a Public Good. *Journal of Personality Social Psychology* 51:110–116.

———— 1988. The Provision of a Sanctioning System in the United States and Japan. *Social Psychology Quarterly* 51(3):265–271.

Zajonc, R. B., et al. 1987. Convergence in the Physical Appearance of Spouses. *Motivation and Emotion* 11(4):335–346.

Index

Abraham (Biblical figure), 77
Abraham, Spencer, 197
accent, 72, 176, 179
acculturation, 88, 179, 190
Aché (Paraguay), 38, 142, 159
action strategies for reputation, 60–61
adopted children, 46, 187–88, 187, 203, 238n6
affines (non-blood relatives), 93–94, 96
African Americans, 84. *See also* blacks
Alexander, Richard, 232n23
Alpesteguia, Jose, 19
altruism, 27–30, *29*, 35, 159, 171, 209, 238n4
altruistic gene, 215–17
altruistic punishment, 39, 60, 70, 236n7. *See also* third-party punishment
Amish communities, 207
ancestral environments, 9–10, 16, 31, 47, 69, 233n35
of Chaldeans, 75–78, *79*, 96–97, 234nn2, 3
anthropology, 65, 68–69, 159, 235n1, 236n4(Ch7)
antifertility agents, 207
Arab American and Chaldean Council (ACC), 87
Arabic language, 78, 182–85, 238n3(AppB)
Arabs, 76, 78, 132, 180, 184–85, 187, 197, 226
Aramaic, 75, 77–78, 182, 238n3(AppB)
archaeology, 68
artificial insemination, 231n13
assessment strategies for reputation, 60–61
assimilation, 88, 203
Associated Food Dealers (AFD), 84
Assyrians, 190, 238n7

astral religion, 77
attentional biases, 73
aunts, 92, *93*, 94–95, 101, 191
autism, 206
Axelrod, Robert, 49–50, 231n16
Aziz, Tareq, 76

Baath Party (Iraq), 76
Babylon, 76–77, 234n3
Baghdad (Iraq), 78, *79*, 80, 184
Baldwin, James Mark, 4–5, 229n1
Baron, Robert, 19, 24
Bell, Paul A., 27
benefits, 37–39
and Core Dilemma, 42–43, 215–16
and direct reciprocity, 48, 50, 57, *58*, 111–12, 116
and ethnicity, 178
and indirect reciprocity, 59, 115
in Chaldean community, 117, 130
and intelligence hypothesis, 41–42
and kinship, 44–45, 231n11
in Chaldean community, 97
and social norms, 66–68, 134, 143–44, 236nn5, 7
in Chaldean community, 146, 148
Big Mistake Hypothesis, 107
and kinship, 89, 91–97, *93*, 235n3
biology, 7–9, 32, 39–40, 177
blacks, 124, 145, 147, 187. *See also* African Americans
blood donation, 30
blood relatives. *See* common descent; kinship
blowguns, 22–23, 232n20
Blurton Jones, Nicholas, 159

255